1557798-1
10/1/03

ONE HUNDRED YEARS OF CHROMOSOME RESEARCH
AND WHAT REMAINS TO BE LEARNED

COVER ILLUSTRATIONS

Left –   Human chromosome 1, from the map published by G.D. Schuler et al. 1996. Science vol. 274: 540–546, 1996.

Center –   First meiotic division in pollen mother cell of *Lilium*. D.M. Mottier 1903 (From F. Schrader, Mitosis, 1944).

Right –   Sequence of events – from Mendel's discovery of the existence of genes, to 2002 and the projected date for completion of sequencing the human genome. P. Little, The Book of Genes, Nature vol. 402: 467, 1999.

BACK ILLUSTRATIONS

Left –   The DNA of *Muntiacus muntjak* cleaved with 7 restriction enzymes and hybridized with read deer DNA. A. Lima-de Faria et al. – DNA cloning and hybridization in deer species supporting the chromosome field theory, BioSystems vol. 19: 185–212, 1986.

Right –   Electron micrograph of a T2 bacteriophage showing the DNA outside the head. A.K. Kleinschmidt et al. – Darsteelung und längenmessungen des gesamten DNA-Inhaltes von T2-Bacteriophagen, Biochem. Biophys. Acta vol. 61: 857, 1962.

Address of Author:
    Department of Cell and Organism Biology
    Genetics Building
    Lund University   Phone int +46 46 2228646
    Sölvegatan 29   Fax int. +46 46 147874
    SE-223 62 LUND   E-mail antonio.lima-de-faria@lu.se
    SWEDEN

ONE HUNDRED YEARS OF CHROMOSOME RESEARCH
AND
WHAT REMAINS TO BE LEARNED

BY

A. LIMA-DE-FARIA

Professor Emeritus
of Molecular Cytogenetics
at Lund University,
Lund, Sweden

KLUWER ACADEMIC PUBLISHERS
DORDRECHT / BOSTON / LONDON

A C.I.P. Catalogue record for this book is available from the Library of Congress.

ISBN 1-4020-1439-2

Published by Kluwer Academic Publishers,
P.O. Box 17, 3300 AA Dordrecht, The Netherlands.

Sold and distributed in North, Central and South America
by Kluwer Academic Publishers,
101 Philip Drive, Norwell, MA 02061, U.S.A.

In all other countries, sold and distributed
by Kluwer Academic Publishers,
P.O. Box 322, 3300 AH Dordrecht, The Netherlands.

Printed on acid-free paper

All Rights Reserved
© 2003 Kluwer Academic Publishers,
No part of this work may be reproduced, stored in a retrieval system, or transmitted
in any form or by any means, electronic, mechanical, photocopying, microfilming, recording or otherwise, without
written permission from the Publisher, with the exception of any material supplied specifically for the purpose of
being entered and executed on a computer system, for exclusive use by the purchaser of the work.

Printed in the Netherlands.

Contents

Introduction   1–2

PART I
NINE PERIODS OF CHROMOSOME RESEARCH: 1795 TO 2010

First Period 1795–1850   5–7
The precursors: Better microscopes allowed reaching the cell level

Second Period 1850–1900   9–12
The pioneers: The discovery of the chromosome was a by-product of microbiology

Third Period 1900–1930   13–15
The era of abstract genetics: Order in embryonic development led to the finding of order in inheritance

Fourth Period 1930–1950   17–20
The impact of physics and chemistry on genetics: World War II encouraged the development of microbial genetics

Fifth Period 1950–1970   21–25
Radioisotopes and electron microscopy became a most fruitful combination: Molecular biology received its main impulse from disciplines outside genetics

Sixth Period 1970–1980   27–30
The mechanisms of cancer and of development were sought at the DNA level: Biotechnology emerged as a new field as genetics created its own weapons

Seventh Period 1980–1990   31–34
Neurobiology reached the molecular level: Artificial chromosomes and gene therapy became a reality

Eighth Period 1990–2001   35–37
The genome of humans and of other organisms was sequenced: The age of multilaboratory collaboration was established

Ninth Period 2001–2010   39–43
The post-genome era: The task that lies ahead

PART II
THE TECHNOLOGY THAT ALLOWED THE STUDY OF THE CHROMOSOME: 1900 TO 2001

From staining methods to DNA sequencing   47–56

PART III
IN SEARCH OF THE EUKARYOTIC CHROMOSOME

Main stages in the discovery of the cell's structure and function   59–62

The nucleus versus the cytoplasm. Which was most important?   63–64

The description of cell division: An impressive transformation was accompanied by directed cellular movements — 65–66

Meiosis was another unexpected property: The cell could reduce its chromosome number — 67–70

The maintenance of identity of the chromosome during interphase was accompanied by constancy and variability of pattern in different tissues — 71–72

PART IV
THE THREE UNIQUE REGIONS OF THE EUKARYOTIC CHROMOSOME

The centromere: A Pandora's box of unearthed properties — 75–79

The telomere: Not just a terminus station — 81–84

The nucleolus organizer: Nothing in the cell is comparable to it — 85–88

PART V
NO CHROMOSOME CAN FUNCTION OUTSIDE A CELL

Cytoskeleton: A disgusting artifact became an important cell edifice — 91–92

Nuclear envelope: The nucleus disclosed its outer structure — 93–94

Centriole: An enigmatic cell invention — 95–96

Endoplasmic reticulum and Golgi apparatus: The building of membranes permitted molecular reactions to occur in defined sequences — 97–98

Cell membrane and cell wall: The cell became an individualized entity — 99–101

PART VI
SPECIFIC TYPES OF CHROMOSOMES

Chromosomes of viruses: An early or a late form of chromosome? — 105–107

Chromosomes of bacteria: Nearly naked DNA could become independent — 109–111

Chromosomes of mitochondria: Intruders invaded the cell — 113–115

Chromosomes of chloroplasts: Additional genomes entered the cell — 117–119

PART VII
THE ANTITHETICAL PROPERTIES OF THE CHROMOSOME

Physico-chemical processes are antithetical — 123

THE CHROMOSOME'S RIGIDITY

Maintenance of organization: The protozoan versus the human chromosome — 127–131

Maintenance of the chromosome phenotype — 133–136

Maintenance of gene order — 137–140

Maintenance of function	141–143
The periodicity of chromosome transformations	145–148

THE CHROMOSOME'S PLASTICITY

Structural change	151–153
Change of pattern	155–156
Change in size	157–158
Change in number	159–160
Change in function	161–165

PART VIII
CHROMOSOME MODELS AND WHAT THEY DO NOT TELL US

The models	169–173
What the models do not tell us	175–176

PART IX
EPILOGUE

Where did the chromosome come from?	179–181
Where is the chromosome going?	183–184
Bibliography: A list of selected books that have dealt with the chromosome during the period 1870–2001	185–189
References: Cited works between 1990 and 2001	191–194
Sources of Illustrations	195–

Subject index

Author inded

Introduction

It may seem inappropriate to write a book about chromosome research during the last 100 years, when one has just entered the period of post-sequencing DNA. A bird's eye view of the chromosome, throughout a century of extensive research, may seem superfluous or irrelevant.

What possible advantage may, present day molecular biologists, draw from the knowledge of previous results or concepts? At first sight it seems that they could easily dispense with it. But some examples may suffice to illustrate how this attitude may not be advantageous. Firstly, in the planning of novel experiments it may turn out to be valuable to know that the gene has been defined in at least 11 different ways since the term was coined, and that the present delimitation and definition continue to be partly elusive. Secondly, it may well be relevant, in approaching future problems, to know which properties have been ascribed to the main chromosome regions and how these have changed throughout the different periods of cell research and continue to do so at present. Thirdly, a sample of the many unanswered questions, which await clarification in the near future, may allow a better formulation of new research programs.

It is true that such an earlier knowledge of chromosome research is not of immediate need if one wants to pursue some of the standard lines of research in biotechnology or to follow conventional areas in molecular cell biology. The cloning of humans has become a standard procedure and isolated genes can easily be transferred from fishes to strawberries without especial additional knowledge. But there may lie the main pitfall. Failure is still the norm in the cloning of humans and other mammalian species – 95% to 97% of the efforts still end in disaster. The transfer of a gene from an animal to a plant, or the reverse process, still poses many unanswered questions concerning the ways in which the newly implanted DNA sequences will behave in relation to the constellation of other genes in the new genome, as well as their ability to be transferred to other organisms. The question is even more acute in the case of human oncogenes and suppressor genes. How are we going to predict their behaviour, their interactions, and their changes of location, without any knowledge of the earlier data on chromosome organization?

Just as it has happened during the development of other sciences, the first century of chromosome investigation led to the abandonment of several concepts and the superseding of earlier dogmas.

Disturbing is also the finding that many phenomena, which have been known since the 1930s, have been shelved or ignored on purpose, to avoid disturbing a neat concept of the chromosome. Although there was a great strength in this way of thinking, because it allowed moving rapidly into the next approach, it left a sea of unchartered islands of knowledge, that have resulted in our present picture of the chromosome as a patchwork of unconcerted phenomena.

This is why this work is not solely about what was achieved during the fruitful previous century, but it is an attempt to find out what remains to be learned about the chromosome. If the past has any interest, it is only to the extent that it may help to charter the future better.

It may be pointed out that this is not a historic work or an Atlas of chromosome research. Instead it tries to furnish a record of observations, experiments and ideas that led to the recognition of the chromosome as the key cellular organelle. The emphasis has been put mainly, not on scientists, but on the results that opened new fields of endeavour. This means that injustices were committed. Many of the leading colleagues, whom I have known personally, if alive today, would stand up and point out some observation or experiment, cherished to them, that I have not included. This is why the figures in the plates represent a difficult choice. In every case, an example of the earliest findings is opposed to the last data available. This was carried out in an attempt to present a trajectory of the development during the last 100 years.

A few comments on the literature cited are necessary. The references given are not accompanied by a detailed bibliography for two reasons. Firstly, such a "Literature Cited" that would cover the 100 years of chromosome research, if only partly, would have too large dimensions. Other publications are available that cover this information. Secondly, the present easy access to computers allows us to find the actual publications without difficulty. However, to simplify the task of the reader two types of references have been included: (1) A list of selected books, that have dealt with the chromosome during the period 1870–2001. These contain detailed references to most of the works cited in the text. (2) Since the papers between 1990 and 2001 may be harder to get from these books, an additional bibliography covering this period has also been included which contains these references in full detail.

Another aspect to be mentioned is that the figures that illustrate a given technique or experiment, were selected, not necessarily from the original work, but from a later publication that depicted them in a simpler way.

The plates were prepared by assembling reproductions of drawings, diagrams, tables and photographs from the available publications. Original photographs cannot be easily obtained since the book covers 100 years of research. However, an effort was made to include the best and most distinct pictures.

What has been achieved so far may be called chromosome prehistory. We now enter a different phase in which we will be in a position to get more precise answers about the many molecular mechanisms which channel the function of the chromosome and which direct its interaction with the rest of the atomic edifice of the cell.

Acknowledgements

This work was supported by research grants from the Swedish Natural Science Research Council, the Crafoord Foundation, the Erik Philip-Sörensens Stiftelse and the Royal Physiographic Society. Thanks are also due to Med. Kand. J. Essen-Möller who kindly contributed with excellent computer work.

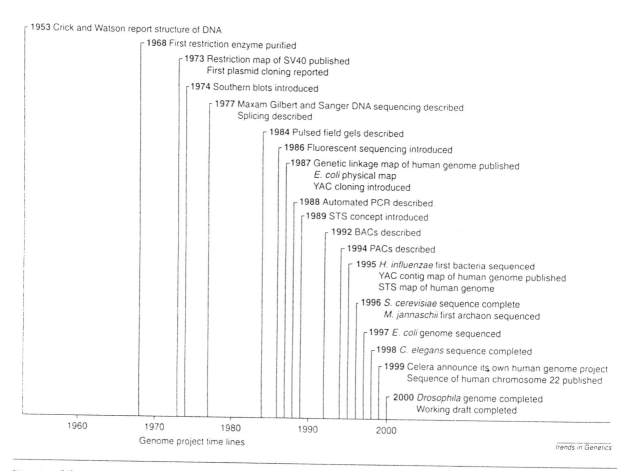

DUNHAM 2000 -
Table describing some of the results, published between 1953 and 2000, that led to the sequencing of the genome of various organisms including humans.

PART I

**NINE PERIODS OF CHROMOSOME RESEARCH:
1795 TO 2010**

First Period 1795–1850

```
             FRENCH REVOLUTION – 1789
                METRIC SYSTEM – 1795
         COMPARATIVE ANATOMY – CUVIER – 1799
        FIRST ELECTRICAL BATTERY – VOLTA – 1800
       TERM BIOLOGY IS COINED – TREVIRANUS – 1802
        PHILOSOPHIE ZOOLOGIQUE – LAMARCK – 1809
    HISTOIRE NATURELLE DES ANIMAUX – LAMARCK – 1815–1822
           SYNTHESIS OF UREA – WÖHLER – 1828
          FIRST PHOTOGRAPHY – DAGUERRE – 1838
         TERM PROTEIN IS COINED – BERZELIUS – 1838
  PIONEERING WORK ON THE CELL – SCHLEIDEN AND SCHWANN – 1839
         COMPOUND MICROSCOPE BUILT AT JENA – 1847
```

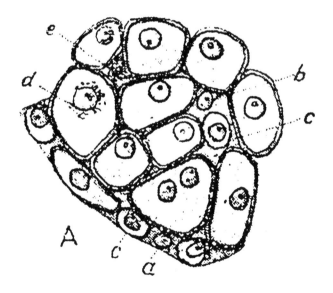

SCHWANN 1839 - Drawing of cells dissected from animal tissue, as seen with the help of the compound microscope, showing the general occurrence of the nucleus and the nucleolus.

The precursors
Better microscopes allowed reaching the cell level

Three major components seem to condition scientific research: 1) No discovery in science arises from nowhere. 2) No result is independent of its historic and socio-economic period. 3) Every discovery is anchored on the technology available at a given time.

Firstly, a discovery, *i.e.* what becomes self-evident to a scientist, is the product of a long series of earlier efforts made by colleagues who have previously worked along similar lines. His result arises as a terminal event built on the labor of many others.

Secondly, the discovery is conditioned by the historic environment in which the scientist finds himself. It is in periods of intellectual upheaval that science best flourishes. The finding of new solutions is aided by intellectual debate and confrontation. The socio-economic conditions are also critical. Science, like the arts, demands a surplus of riches. Affluent societies are able to sustain artists and scientists. These hold professions that under other conditions would be considered superfluous or luxurious.

Thirdly, the technology that is available at a certain time, is critical to the realization of a given experiment or to the concretizing of an idea. It is the instrument available that puts limits to what can be tested with accuracy.

The discovery and study of the chromosome are a good example of the importance of these three components.

It may sound strange, that to understand the establishment of the concept of the chromosome, which occurred at the transition stage between the 19th and 20th centuries, one needs to go back to the whole of the 1800s. Actually, it is also appropriate to consider the Encyclopaedists of the middle of the 18th century, for it is they who paved the way to the intellectual fermentation that culminated in the French Revolution. Even a brief mention of the achievements of the 16th century is in order since they were the direct cause of the birth of modern science.

Science is a fragile process that only grows well when obscurantism is exposed and pushed aside. It may be recalled that the society of the Middle Ages was highly static and irrational. It was the discoveries of the new sea routes and new continents, carried out mainly by the Portuguese and the Spaniards, in the 1400s and 1500s, that shattered the Middle Age society. New human beings, new animals, new plants, arrived every month in Europe breaking old preconceived ideas. Moreover, the Earth was proved to be a sphere by circumnavigating it. An immediate consequence of these findings was an explosive expansion of trade.

The empiric approach used on the chartering of the oceans, led to an equally empiric method on land that resulted in a new way of investigation and a renaissance in science. The human body was now a source of permanent discovery and could be freely dissected to its smallest details. The brain could be shown to be connected by nerves to all the other organs of the body. The human blood was found to circulate in a predictable way. The newly invented microscope permitted discovering a new world of organisms that could not be seen previously with the naked eye. And the use of the telescope led to the finding that Jupiter had moons that orbited around it, like the Moon circling around our Earth.

But at every age, just like today, science was being threatened. Established interests charged against the new intellectual endeavour taking the form of religious wars that devastated Europe in the 17th century. This is why the latter work of the Encyclopaedists was so important. Between 1751 and 1780 they published 35 volumes compiling the technology and science of the age.

It is in this atmosphere that a critical technological event occurred, that turned out to be of basic significance for the future development of the different branches of science. Until then, every country, even every city, had its own measurements and weights. This situation had many disadvantages. Firstly, the units were not well defined. Secondly, they could not be compared with one another with accuracy.

Within months of the establishment of the French Revolution, the president of the new Assembly, C.M. Talleyrand, proposed in 1790, the establishment of a commission, within the Academy of Sciences, to create the meter and the kilogram. These units had for the first time a scientific basis. The meter was a fraction of the Earth's meridian from the North Pole to the Equator passing through Paris. The kilogram was based on a cube of water at 4 °C, the temperature of the water's maximum density. Science is essentially based on measurement. The impact of the new units turned out to be invaluable.

The period 1795–1850 was characterized by the pioneering work on living organisms and on the cell.

The knowledge of the human organs was extended to that of most animals, which resulted in the unified science of comparative anatomy established by G.L.C. Cuvier (1799). But even animals that no longer existed could be studied. It was also G.L.C. Cuvier who laid the foundations of animal paleontology. The fact that the same basic pattern recurred in all vertebrates and that their organs were related, led G.L.C. Cuvier to predict the whole pattern of the skeleton of an animal from a single bone. His principle of the correlation of parts was a breakthrough in the life sciences. It demonstrated the order underlying the construction of living organisms. It also showed that one could predict the whole from its parts, the ability to predict being one of the pillars of scientific endeavour.

The work of Georges de Buffon (1707–1788) led him to describe in detail many unknown organisms. This first general view of the living world was concretized in 44 volumes published between 1749 and 1804.

It is thus not surprising that G.R. Treviranus coined the term biology (1802) that was soon taken up by J.B.M. Lamarck in his work "Philosophie Zoologique" (1809) and in his extensive treatises on the "Histoire Naturelle des Animaux" (1815–1822).

J.B.M. Lamarck's main contribution is based on his courage to refute the independent origin of animals stated in the Bible. A. Lavoisier also had the courage to expose the concept of the phlogiston, that had prevailed in chemistry as a dogma for nearly 100 years. Both could do it thanks to the breaking of the intellectual codes brought about by the new social transformation.

For J.B.M. Lamarck living organisms were not accidental structures, disconnected and unrelated, but could be ordered from the simplest to the most complex, an order that was based on their biological affiliation. The actual mechanism responsible for the transformation of organisms could obviously not be available since the phenomenon was just being discovered. His work received additional support from E. Geoffroy Saint-Hilaire, a zoologist, at the Museum of Natural History in Paris. He wrote two volumes in 1818–1822 emphasizing the "unity of composition" according to which there was a structural basic plan common to all vertebrate animals.

A technical development was bringing biology closer to the cell level. The compound microscope that was built already in the 16th century was improved by the inclusion of better lenses in 1733 by C.M. Hall, and the Zeiss optical works opened at Jena, Germany, as early as 1847. Not only organs but tissues and their ultimate components, the cells, started to become the center of attention by using the new microscopes.

Simultaneously chemistry was developing rapidly. J. Dalton formulated his atom theory nearly at the same time as the theory of biological evolution was proposed (1810). Nothing was known at that time of the composition of atoms, but it was the starting point of the study of their internal structure.

An epoch-making discovery in chemistry was F. Wöhler's synthesis of urea (1828). He could demonstrate for the first time that an organic substance could be synthesized in the laboratory. This was a heavy blow to the teleological idea of a "vital force" which was supposed to characterize the chemical processes occurring in living organisms. From now on the molecules found in cells could be synthesized in a glass vial and in this way could be shown to be indistinguishable from those being formed outside the cell.

Soon after, J.J. Berzelius's discoveries in chemistry led him to coin the term protein (1838). This was also the time when the first enzyme was identified (diastase).

Another type of technology that sprung from chemistry helped in the identification of the cell components. The use of dyes led to the localization of cellular organelles, such as chloroplasts, nuclei and nucleoli, as well as to the localization of cellular compounds. Histochemistry was born when F.V. Raspail employed iodine to recognize the occurrence of starch in cells (1825). Another application of chemistry led L.J.M. Daguerre (1838) to produce the first photographs using silver salts. In the future an exact documentation of a structure, or of a phenomenon, could be obtained and preserved, without the bias of the observer. In physics A.H.L. Fizeau (1849) measured the speed of light, showing that it was not infinite but amenable to physical evaluation.

Some of these disciplines converged into the establishment of what has been called the theory of the cell, established by M.J. Schleiden and T. Schwann (1839). They showed that plant and animal cells were essentially alike, a finding that confirmed the basic similarity of living organisms at the microscopic level. However, the mechanism by which cells originated remained obscure. Soon, by 1844, A. Kölliker discovered the cell multiplication process and H. von Mohl (1846) created the word protoplasm, to emphasize the fundamental nature of the living material.

This period closed with an experiment (1850) which attracted great interest among the population of the city of Paris. J.B.L. Foucault constructed a pendulum hanging from the ceiling of the hall of the Pantheon. Its movement along a changing pass demonstrated the rotation of the Earth. This result made the Earth a body, like any other, rotating in outer space.

The major events of the period were: 1) The combination of the anatomy of living organisms with that of fossils. 2) The establishment of the concept of evolution. 3) The unity of cell construction. 4) The synthesis of organic compounds in the laboratory. What had previously been considered disparate and unrelated suddenly became unified into a common denominator of structure, function, descent and composition. Where before there was chaos, order and unity started to be unveiled.

Second Period 1850–1900

EPIDEMICS OF CHOLERA, SYPHILIS, TUBERCULOSIS, LEPROSY AND OTHER DISEASES.
SEARCH FOR MICROBES
VACCINATION AGAINST SMALLPOX – 1853
NEW CELLS AROSE ONLY FROM PREVIOUS ONES – VIRCHOW – 1855
FERMENTATION CAUSED BY LIVING ORGANISMS – PASTEUR – 1857
ORIGIN OF SPECIES – DARWIN – 1859
INHERITANCE IN PEAS – MENDEL – 1866
ORDER IN EMBRYOLOGY, HUMAN EGG – VON BAER – 1828, HAECKEL – 1866
PERIODIC TABLE OF ELEMENTS – MENDELEEV – 1869
CELL DIVISION IN PLANTS – STRASBURGER – 1875
BACTERIOLOGY AS A SCIENCE, STAINED PREPARATIONS – KOCH – 1882
LONGITUDINAL HALVES OF CHROMOSOMES SEPARATE AND MOVE TO ENSUING
DAUGHTER CELLS – VAN BENEDEN – 1883
LAWS OF CHROMOSOME BEHAVIOUR – VAN BENEDEN – 1884
TERM CHROMOSOME COINED BY WALDEYER – 1888
X-RAYS – ROENTGEN – 1895
RADIOACTIVITY – BECQUEREL – 1896
GOLGI APPARATUS – GOLGI – 1898

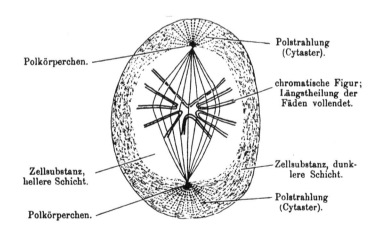

WALDEYER 1888 -
In his work on cell division he introduced the term "chromosome" but in this figure he still called them "filaments" (Fäden).

The pioneers
The discovery of the chromosome was a by-product of microbiology

Most cytologists do not seem to have realized that the discovery of the chromosome was mainly a by-product of the active search for the microbes that were found to cause the diseases that afflicted humanity during the second half of the 19th century.

Cholera, syphilis, tuberculosis, typhoid fever, leprosy, rabies, malaria and other diseases were epidemic and led to a dreadful harvest of lives of all ages. Their cure became an imperative. It initiated a search for the microbes that were suspected to be their agents. At the same time an effort was being made to counteract these diseases chemically.

In a first step vaccination against smallpox was made compulsory (1853). It is, however, the development of bacteriology, initiated and established on a solid scientific basis, by L. Pasteur, that had a strong impact on the biology of the cell. He started by analysing the fermentation of wine which turned out to be caused by living organisms that were microscopic (1857). Then, he went on to cure the silkworm disease thus saving the French silk industry. Later he observed the action of antibiotics on the anthrax bacillus and still later successfully inoculated a child bitten by a dog infected by rabies. Among the many improvements that he introduced into microbial research he was the first to culture outside the body a microscopic organism that was responsible for disease. He finally proved that microorganisms caused illness.

During his discoveries on microbes L. Pasteur was obliged to look for an answer to one of the main questions of that period. What was the origin of these microbes, or what was actually the origin of life? Did it arise spontaneously every time from nowhere or was it a prisoner of a continuous process from which it could not depart? L. Spallanzani, in Italy, had shown about 100 years earlier (1767) that living organisms, only originated from previous living organisms. But, who could believe that a beautiful bee or a fly could emerge from a totally different animal such as the highly amorphous larva that had no wings, no legs and looked like a worm?

Those who defended the spontaneous generation had no difficulty in pressing their argument in face of such a drastic difference between the two stages of development of the same animal. L. Spallanzani's experiments had been as carefully carried out as those of L. Pasteur, but now there was another mental attitude based on the pressing need to fight infections. Once L. Pasteur proved that there was no spontaneous generation (1861) the result became generally accepted.

Another leading bacteriologist at the time was H.H.R. Koch, who discovered the tubercle and cholera bacillus. During the same period E. Metchnikoff proposed that white blood cells (phagocytes) destroyed bacteria and R. Ross discovered that the cause of malaria was a microbe (1897).

Everywhere there was a hunt for microbes, outside the body, in the body, in the blood, in the cell. R. Virchow was an outstanding pathologist, who suggested that the cell was the prime point of disease, a revolutionary thought at a time when so little was known about this living unit. It was he who argued that all cells can arise only from preexisting cells which was stated in the aphorism: "Omnis cellula e cellula", (1855). This statement is often attributed to him but was actually coined by F.V. Raspail in 1825. R. Virchow in his book on cellular pathology (1858) emphasized the importance of this concept contributing to establishing it as a general principle.

L. Pasteur's and R. Virchow's discoveries vindicated the notion of life's continuity both at the organism and at the cell levels.

The stage was set for social confrontation, since this was a convulsive social period in which rapid industrial expansion dealt with a restive working class. Some of these scientists were actually involved in the revolutionary movement. R. Virchow fought on the barricades in Berlin, during the uprising of 1848, the result being that he was dismissed from his scientific position.

Moreover, slavery and serfdom were still a general feature not only in the colonized territories, but in large parts of Europe. These forms of human oppression were formally abolished in Russia in 1861 and in the U.S.A. in 1863, but it took a long time to extinguish their effects.

Three major works reflected this situation: T.R. Malthus's "An Essay on the Principle of Population" (1798), the work of J.A. Gobineau "Essai sur l'Inégalité des Races Humaines" (1853), followed by the book of F. Galton "Hereditary Genius" (1869). All three extolled class struggle and hereditary supremacy. As C. Darwin acknowledged, his interpretation of evolution was based on T.R. Malthus's essay on population. C. Darwin's contribution to evolution was not only the "Origin of Species" (1859), as is usually assumed, but equally important was his work on the "Descent of Man" (1871). Humans became definitely biologically anchored to other animal species as a result of the comparison of anatomical and behavioristic traits. At the time C. Darwin wrote both works, heredity was not understood. The result was that he based evolution on two main processes: the inheritance of acquired characters and selection. As E.B. Wilson pointed out in his classic book "The Cell" (1925): "Darwin thus assumed (in accordance with a notion even now widely prevalent) that the parent literally transmits its characters to the offspring, and thus sought to explain the heredity of "acquired" or "somatogenic" characters, at that time generally accepted as a fact". One had to await the discovery of mutations in *Drosophila* to establish neo-Darwinism on a firm basis.

Interestingly Darwin satisfied both sides of the social conflict. On one hand, he established evolution on a broader scientific basis, mainly supported by the newly discovered stratigraphy of rocks and by a vast array of recent biological information. This led to an attack from the Church and put him on the side of the progressive forces. On the other hand, by introducing selection as an agent of evolution, he gave biological credentials to those who oppressed the underprivileged and defended the expansion of a "laissez faire" economy.

This view of the path of biological and human evolution led to a long debate that disregarded fundamental results that had become available in other fields of biology. This appears to be one of the main reasons why G. Mendel's work on the inheritance in plants, published seven years later (1866) was not taken into consideration. It offered the first demonstration of a predictable order prevailing in hereditary transmission. The same treatment was given to the embryological work of K.E. von Baer and of E.H. Haeckel. A causal relation was established by E.H. Haeckel between embryology and phylogeny. This was better elaborated by K.E. von Baer (1828) who enunciated four laws concerning embryonic development. Nothing appeared more intrinsically determined than embryonic development. On the other hand, the struggle for survival, the prime mover of selection, implied an indiscriminate event, occurring at the organism's level, that left little place for intrinsic processes.

The prevalence of this last attitude also extended to other sciences, such as chemistry.

Since the beginning of the 19th century an increasing number of chemical elements had been identified and their properties did not seem to occur at random. But, who would even dare to suggest that there was order at such a basic level as that of the chemical elements? In England, J.A.R. Newlands considered such a suggestion, as early as six years after the "Origin of Species", and exactly in the same year that G. Mendel published his rules of inheritance. J.A.R. Newlands proposed the "law of octaves" based on the similarities of the basic properties of the elements. His paper was rejected by "Nature" and ridiculed. No one could accept any order in chemistry. However, in Russia, far away from the English turmoil, D.I. Mendeleev could pursue his chemical studies and formulate in 1869 his law according to which "the properties of the elements are a periodic function of their atomic weights". He published a table containing 63 chemical elements and was able to predict the properties of 10 as yet unknown. These were discovered soon and were found to have the properties predicted by D.I. Mendeleev. Chemistry was becoming an exact science.

In the meantime, in the silence of the laboratories, far from the scientific and political debates of the day, others continued to look deeper into the cell's organization. Many microorganisms appeared as filaments in body fluids and in cells. Hence, it was necessary to find a method to distinguish the microbes from permanent cell organelles. It is at this stage that the chromosome, is revealed as a result of the bacteriological analysis, and it assumes the position of a significant component of the cell.

By 1882 W. Flemming had discovered lampbrush bodies in oocytes and had coined the term mitosis for the cell division. In the next year E. van Beneden studied *Ascaris* cells that contained only four filaments. He showed that the gametes contained only two of them and that the number of four was reestablished following fertilization. In the same year (1883) W. Roux drew special attention to the "filaments" present in the cells during division and W. Waldeyer called them "chromosomes" in 1888.

The term chromo + soma derives from the fact that the technology that had been so successfully developed in bacteriology for staining microbes turned out to be equally effective in dyeing these filaments that were permanent components of the cell. Some years earlier L. Ranvier had introduced the "picro-carmin" technique, and by 1870 other staining methods had been added to microscopy.

The studies of W. Flemming (1843–1915) and of E. Strasburger (1844–1912) led to two main findings which opened the door to a deeper knowledge of the chromosome: 1) The nucleus was the organelle of the cell that directed the division process. 2) Mitosis was a phenomenon with features common to animals and plants.

The final proof came when E. van Beneden (1883) found that, in both plants and animals, the longitudinal halves of each split chromosome separated from each other and that they were transferred to the two ensuing daughter nuclei.

For many years one has referred to G. Mendel's results on the inheritance of characters in peas as G. Mendel's laws. Equally important, but not usually referred to in this way, were E. van Beneden's laws (1883–1884). They stated: 1) The chromosome number that appears in the first cleavage of the zygote is always the sum of the haploid numbers received by the gamete nuclei in meiosis. 2) In hybrids between individuals with different chromosome numbers the zygotic number is always equal to the sum of the two parental haploid numbers. This was demonstration that the chromosomes behaved as independent units and that they did not lose their identity from generation to generation. This result actually represented a parallel event to G. Mendel's segregating factors. There was continuity and independence both for the organism's characters and for chromosome transfer. This was the first critical step leading to the future synthesis between genetics and cytology.

The 19th century ended with a series of impressive achievements.

A new organelle was discovered inside the cell. Thanks to the use of a new histological technique, that he had developed, and that employed silver nitrate to stain nerve cells, C. Golgi (1898) observed a structure that bears his name today. In the beginning considered an artefact one had to await the advent of the use of the electron microscope in the 1960s to establish its exact structure and function.

But it was physics that filled the turn of the century with four basic results: W.C. Roentgen discovered X-rays (1895), Antoine H. Becquerel radioactivity (1896), J.J. Thomson the electron (1897) and finally Marie and Pierre Curie isolated radium and polonium (1898). Humanity had suddenly entered into a new age that it was not prepared to cope with.

Third Period 1900–1930

REDISCOVERY OF MENDEL'S LAWS – CORRENS, TSCHERMAK, DE VRIES – 1900
TERM MUTATION COINED BY DE VRIES – 1901
SEX CHROMOSOMES – McCLUNG – 1902
TERM GENETICS COINED BY BATESON – 1903
CROSSES IN ANIMALS FOLLOW MENDEL'S LAWS – CUÉNOT – 1905
TERM GENE COINED BY JOHANNSEN – 1909
DROSOPHILA BECOMES A GENETIC TOOL – MORGAN – 1910
FIRST MAP OF DROSOPHILA – STURTEVANT – 1913
INDUCTION OF MUTATION BY X-RAYS – MULLER – 1927
STAINING COMBINED WITH SQUASH TECHNIQUE – BELLING – 1927
DISCOVERY OF HETEROCHROMATIN – HEITZ – 1928
CHIASMATA FORMATION – DARLINGTON – 1929
POPULATION GENETICS – FISHER, HALDANE, WRIGHT – 1930

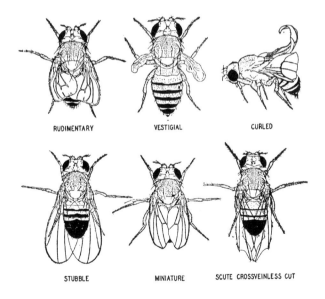

WALLACE 1940 -
Various wing and bristle characters in *Drosophila*.

The era of abstract genetics
Order in embryonic development led to the finding of order in inheritance

The year 1900 is known to geneticists as a turning point in the development of their science. One assumes many times that scientific problems come from nowhere, that the scientist enters his laboratory free of preconceived ideas and that he is searching for what is supposed to be the truth. This is the naive view of those who are not acquainted with the making of science. There is no better example that helps to dispel such an erroneous view than the rediscovery of G. Mendel's laws. In 1900 three different scientists, working in three different laboratories and in three different countries, performed independently of one another the same experiment and obtained the same result. H. de Vries, C. Correns and E. Tschermak carrying out research in Holland, Germany and Austria respectively, rediscovered G. Mendel's "laws" of inheritance. All three entered their laboratories with the definite intention of testing the mechanism of inheritance. What led them to test hybridization among plants and to look for order in descent?

The scientific controversy that had prevailed in many circles between 1850 and 1880 was now dying out. Darwinism was considered inadequate as an explanation of evolution, since there was no clear understanding of the hereditary material on which selection could act. At the same time embryology had moved to a preeminent position in biology. All embryos were related by common structural features that had not changed appreciably during evolution.

An imperative question arose. Was there an ordered mechanism beyond heredity? H. De Vries, C. Correns and E. Tschermak looked for a plant species that could give an appropriate answer.

The result was astounding. From now on it was possible to predict the proportions in which a given character appeared in the progeny following a cross hybridization. In advanced sciences, such as physics, prediction had been a regular event since the establishment of the theorems of Galileo Galilei and the laws of I. Newton, formulated in the 17th century. Biology was now becoming a mature science by displaying the same capacity. The consequences were manifold. H. De Vries discovered a new hereditary process. He coined the term "mutation" for the changes in the organism's characters that were transmitted to the progeny (1901). C.E. McClung suggested that the "accessory chromosomes" found in unequal number in insect sperm were sex determinants (1902). T. Boveri followed the development of sea urchin embryos which were haploid, diploid and aneuploid. Only diploid embryos developed normally, *i.e.* those having the full set of chromosomes. From this result he concluded that the chromosomes were the hereditary determinants of the animal's development (1902). But the discovery of W.S. Sutton was crucial. By analysing the pairing of chromosomes at meiosis and the independent separation of bivalents at anaphase I, he concluded that such a behaviour supported the independent assortment of characters observed following hybridization (1902). This synthesis led to the formulation of the chromosome theory of heredity.

At this time W. Bateson introduced, among other terms, that of "genetics" for the new science (1903). It was soon followed by W. Johannsen's coining of the word "gene" (1909).

The experiments of G. Mendel, and of his followers, had all been carried out with plants. L. Cuénot (1905) showed that crosses in animals led to the same type of inheritance found in the plant kingdom.

Other phenomena started to become evident. These were: 1) The linkage between hereditary traits (1906). 2) The formation of chiasmata between homologous chromosomes (1909). 3) The cytoplasmic inheritance of various traits (1909). 4) The inheritance of quantitative characters. These last two phenomena seemed from the beginning to be exceptional events but became amenable to a Mendelian explanation (1909).

T.H. Morgan, who was originally an embryologist, entered the field of genetics full of criticism. If one was going to know what actually happened at the chromosome level one could not wait one year to obtain results. Hybridization in plants and higher animals was a process that took too long. He looked for an animal that could produce a new generation in 10 to 20 days. *Drosophila* was the choice because it also produced several hundred offspring and could be raised on simple culture media (1910). A whole river of knowledge has flown from the laboratory use of this insect and has not stopped to this day. No living organism is so well known, and has led to collecting so many scientific results, as this small and inconspicuous fly. Most significant is that what has turned out to be true for *Drosophila*, is true for elephants and for humans. What a lesson in humility is gathered from contemplating this minute animal.

The first genetic map of *Drosophila* was put forward by A.H. Sturtevant (1913). T.H. Morgan, A.H. Sturtevant, C.B. Bridges and H.J. Muller, discovered a series of unknown phenomena: sex-linkage, non-disjunction, chromosome deficiencies, balanced lethals, chromosome duplications, monosomics, attached X-chromosomes, chromosome translocations, position effects among genes, chromosome inversions and other phenomena. A new experimental period was opened when H.J. Muller reported the induction of mutations in *Drosophila* by X-rays (1927). These mutations were studied in detail by many workers including E.M. Wallace (1940).

The work on *Drosophila* was accompanied by a similar analysis of chromosome behaviour in different species of plants. The *Datura* work of A.F. Blakeslee and his associates resulted in an elegant demonstration of the relationship between trisomic plants and phenotypes specific for every chromosome type (1920). J. Belling showed that interchanges between non-homologous chromosomes resulted in the chromosome rings observed at meiosis (1927) a result confirmed in Oenothera by R.E. Cleland and A.F. Blakeslee (1930). A most startling experimental result was obtained when G.D. Karpechenko created an allotetraploid hybrid of radish (*Raphanus sativa*) and cabbage (*Brassica oleracea*) which he called *Raphanobrassica*. This meant that plants representing separate genera could be united leading to the synthesis of a new organism (1927).

The advances in staining techniques and in histochemistry continued. R. Feulgen and H. Rossenbeck described the test which allowed localizing DNA in chromosomes, distinguishing it from RNA and other cell molecules (1923). J. Belling made his detailed studies of chromomere patterns at the prophase of meiosis in several plant species. He also improved the staining techniques by introducing the acetocarmine method combined with the squash technique (1926). Most important, but of no immediate significance at the time for cell biology, was T. Svedberg's construction of the first ultracentrifuge (1923) which would later turn out to be invaluable in the analysis of proteins and DNA.

As the period closes three major contributions herald the coming of the next interval. These are E. Heitz's discovery of heterochromatin (1928), C.D. Darlington's work on chiasmata formation and movement (1929–1931) and the mathematical analysis of population genetics by R.A. Fisher, J.B.S. Haldane and S. Wright (1930–1932).

This was a most fruitful time which resulted in the consolidation of genetics transforming it into an independent science.

What characterized this period was the establishment of a firm relationship between chromosome properties and hereditary events. Practically any new exception or deviation found in the inheritance of characters could be assigned to an exact deviation in chromosome behaviour.

However, this period has been called by some geneticists, the era of abstract genetics. The reason is that during this time the chemical nature of the gene and the chemical nature of most cell organelles remained unknown.

Biologists tend to think in terms of high complexity associated with cellular events. Following this line of thought the gene could not consist of the simple nucleic acids that were known to be constituents of chromosomes. The gene, the bearer of such a complex process as heredity, had to be complex itself, and for this reason had to consist of large molecules such as proteins which were abundant in the chromosome and in the cell. This is an assumption that persisted even through the 1930–1950 interval.

In the meantime a socio-economic catastrophe occurred. The collapse of the New York Stock Exchange (1929) led to a global economic crisis. This event combined with military interventions put Europe in a state of alarm that presaged the Second World War (1939). Such a state of affairs had a decisive impact on the direction of research mainly in physics and genetics.

Fourth Period 1930–1950

FIRST ELECTRON MICROSCOPE – KNOLL AND RUSKA – 1932
PROTEIN ELECTROPHORESIS – TISELIUS – 1933
GIANT POLYTENE CHROMOSOMES – PAINTER, HEITZ, BAUER – 1933
NUCLEOLUS ORGANIZER – McCLINTOCK – 1934
PHASE CONTRAST MICROSCOPE – ZERNICKE – 1935
NUCLEIC ACIDS IN CELL ORGANELLES – BRACHET – 1936, CASPERSSON – 1936
TOBACCO MOSAIC VIRUS CONTAINS RNA AND PROTEIN – STANLEY – 1937
PHAGE GENETICS – DELBRÜCK – 1939
NEUROSPORA GENETICS – BEADLE AND TATUM – 1941
GENETIC TRANSFORMATION IN BACTERIA – AVERY ET AL. – 1944
DNA AMOUNT CORRELATES WITH CHROMOSOME NUMBER – BOIVIN ET AL. – 1948
STRUCTURE OF CENTROMERE AND ITS DNA CONTENT – LIMA-DE-FARIA – 1949–1950
TRANSPOSABLE ELEMENTS IN MAIZE – McCLINTOCK – 1950
REGULARITIES IN THE BASE COMPOSITION OF DNA – CHARGAFF – 1950
ULTRA-THIN SECTIONS IN ELECTRON MICROSCOPY – 1950

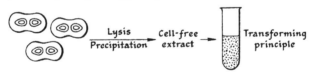
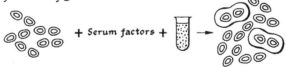

AVERY, MacLEOD and McCARTY 1944 - Experiment demonstrating that DNA was a carrier of hereditary information. Transformation of R mutants of the pneumococcus into S types by a cell-free extract (transforming principle) of S donor bacteria which contained only purified DNA.

The impact of physics and chemistry on genetics
World War II encouraged the development of microbial genetics

Technically this is a remarkable time. Physics started to invade the realm of biology. The first model of the electron microscope was invented by M. Knoll and E. Ruska (1932). This instrument was going to be of crucial importance, during later decades, in the study of cell organelles, viruses and isolated macromolecules.

From the beginning, and during many years, the electron microscope was a disappointment to chromosome researchers. Thin sections of metaphase chromosomes produced no information about its internal organization. The chromosome was simply too thick and too compact to allow a resolution about its internal structure. One had to wait for later methods, in which its fibrillar pattern was disrupted, to obtain better pictures of its contents. Yet, still today, the electron microscope has not revealed much that is fundamentally new about the chromosome's internal pattern.

Another impact of physics into biology resulted in the invention by A.W.K. Tiselius, in the following year, of the technique that allowed separating molecules in an electric field. Electrophoresis would also turn out later to be an invaluable method in the separation of proteins as well as DNA sequences.

A third product of the inroads of physics into biological technology was the discovery of the phase contrast microscope by F. Zernicke (1935), which allowed the study of cells and tissues under living conditions.

However, the great breakthrough took place when physics and chemistry were united in an effort to probe the cell's molecular make-up. T. Caspersson, and independently J. Brachet, used cytospectrophotometric and cytochemical methods to obtain quantitative measurements and to localize: RNA, DNA and protein in cell organelles (1936). Their results allowed distinguishing between the localization of DNA and RNA in the nucleus. Moreover, they made it possible to demonstrate the presence of nucleic acids in organelles associated with the chromosome, such as the nucleolus, which until then had not been defined chemically. The functions of the chromosome and of the cell were now starting to be understood in chemical terms.

Chemistry had an impact of its own. Among other results, A.F. Blakeslee and A.G. Avery reported that colchicine induced polyploidy in plants (1937). This was a simple and effective way of increasing the chromosome number, a phenomenon that had been found to be a natural process in species formation. Later this technique was used effectively in plant breeding.

Even more important was A. Claude's development of the cell fractionation techniques using ultracentrifugation, which was followed by the biochemical characterization of the different components (1946).

In the meantime a great deal was happening at the chromosome level. By producing special combinations of chromosome rearrangements, H.B. Creighton and B. McClintock demonstrated in maize that crossing over actually occurred at the cytological level. A similar proof of this phenomenon was obtained in *Drosophila*, in the same year, by C. Stern (1931).

One unexpected finding was the discovery that the giant structures found in the salivary glands of *Drosophila* larvae were chromosomes. E. Heitz and H. Bauer (1933) and independently T.S. Painter (1933) reported that they were polytene structures, in this way explaining their morphology. These chromosomes had such a gigantic size that in some species they could be seen with the naked eye. Also, the banding pattern that they displayed allowed a degree of precision in the location of genes along the chromosomes, that no one had dreamed of.

It was C.B. Bridges who refined the analysis of the salivary gland chromosomes. In 1935 he published the first complete map of all four *Drosophila melanogaster* chromosomes, in which all bands were numbered and he superimposed the known data from genetic linkage on the cytological map. This was an impressive synthesis of genetic and chromosomal information. Other researchers concentrated on the correlations between structural rearrangements and gene location as the resolution of the banding pattern improved all the time. The use of minor deficiencies allowed a dissection of individual bands. The result was the location of genes in these minute segments, a work carried out by H. Slizynska for the *Notch* locus (1938).

Parallel to the work in *Drosophila*, was the effort made by several colleagues, to make of the maize plant, an equally valuable object of genetic research. To this group belonged B. McClintock who found a translocation which disclosed that the nucleolus organizing region could be divided into two distinct functional components (1934). This meant that specific chromosome segments were multiple structures.

Suddenly DNA became more important than envisaged before. However, its significance continued to be elusive. M. Schlesinger found that some bacteriophages consisted of both protein and DNA (1934). The newly isolated tobacco mosaic virus turned out to contain 5% RNA besides protein (F.C. Bauden and N.W. Pirie 1937). Ultraviolet light was effective in inducing mutations in *Sphaerocarpus* and its action corresponded to the wave length at which nucleic acids had their maximum absorption. The proteins began to lose their position as primary genetic agents, since they were dependent on nucleic acids for their synthesis (1941).

In 1944 O.T. Avery, C.M. MacLeod and M. McCarty performed an experiment which revealed that in bacteria, the agent of genetic transformation was mainly DNA. This result strongly indicated that DNA was the genetic material, but it had little impact on the research carried out in higher organisms. The reasons were several: 1) The results were published in the middle of World War II, when international contacts between many scientists were severed. 2) At that time it was not known that bacteria had chromosomes comparable to those of other organisms, and most genetic thinking was based on the correlation of genetic events with chromosome behaviour. 3) A clear distinction between the roles of RNA and DNA, in the cell of plants and animals, remained to be elucidated. As late as the mid-1950s it was thought that in plants the dominating genetic material was RNA and not DNA. This last nucleic acid, for many years, turned out to be difficult to extract from plant tissues.

The work of O.T. Avery and his collaborators was part of a larger research movement that led to the emergence of a new discipline within biology: microbial genetics. Phages, bacteria and moulds had become an object of inquiry in order to find out their structure, their functions and genetic behaviour.

The interest in microorganisms was not accidental. A great military confrontation was approaching. During World War I (1914–1918) Germany was the country with the most developed chemical industry. The result was the use of poisonous gas as a devastating weapon, in the final phase of the war. Now, in the mid-1930s two other types of new weapons were being actively developed. The discovery of artificial radioactivity in 1933 by F. Joliot and I. Joliot-Curie left no doubt that an atom bomb could be constructed. The question was how quickly it could be produced. If the bomb could not be obtained in time one had to possess another equally devastating agent. The alternative was germ warfare. The study of bacteria and viruses suddenly became a project that received high priority by the 1930s. The tobacco mosaic virus was isolated and crystallized by W.M. Stanley in 1935 and H. Karström studied bacterial enzymes and the substrates that they attacked (1937). E.L. Ellis and M. Delbrück initiated phage genetics in 1939, and bacterial research was transformed into a new discipline of genetics when S.E. Luria and M. Delbrück demonstrated that mutations occurred in these organisms (1943). This was soon followed by the finding that mutations occurred in bacterial viruses and by the discovery that genetic recombination took place both in viruses and in bacteria (1946, 1949).

Another question that then arose was: What were the chemical intermediates between the gene product and the organism's final structure and function?

In a primary effort to answer this problem G.W. Beadle and B. Ephrussi used *Drosophila* to analyse the biochemical pathways participating in eye-pigment synthesis (1935). Similar work was carried out in *Ephestia* by A. Kuhn and A. Butenandt in the same year. This work had the merit of encouraging the research in *Neurospora*, a mould that grows in a well defined culture medium. This allowed carrying out the biochemical analysis of gene products with high precision. The work on this microorganism led to the formulation of the "one gene – one enzyme" hypothesis by G.W. Beadle and E.L. Tatum (1941). By 1948 this hypothesis received confirmation as a result of the work of H.K. Mitchell and J. Lein. In the meantime B. Ephrussi and collaborators showed that yeast was an ideal microorganism for genetic studies (1949). To this day yeast remains one of the best studied organisms from the biochemical and genetic point of view. It is not by accident that this was the first species in which the bases of the DNA of all the chromosomes were fully sequenced (1996).

Another question that started to preoccupy geneticists was the actual size of the gene. How big was it? At that time it was assumed that the gene consisted of protein. A first calculation was made by N.V. Timofeyeff-Ressovsky and M. Delbrück, based on the so-called "target theory", derived from the analysis of mutations, following the action of radiation. Later this problem was taken up by a physicist, E. Schrödinger, who calculated the size of the gene in 1944, reaching the conclusion that it should consist of about one million or a few million atoms. This value was partly based on the estimated size of single bands of salivary gland chromosomes, that C.D. Darlington had calculated to have a volume equal to a cube of 300 Å. The assumption made at that time was that one band corresponded to one gene.

This period closes with some remarkable findings that had no immediate impact but that turned out later to be quite important.

A. Boivin, R. Vendrely and C. Vendrely (1948) measured the quantity of DNA in different tissues and found it to be correlated with the chromosome number. A controversy issued due to the finding of some exceptional cases, mainly in oocytes which were reported to have too much DNA. It turned out later, that both the rule and the exceptions, were based on correct measurements. It is this debate that led to what was called "metabolic DNA" which in turn opened the way to the study of gene amplification, that was just discovered in animal oocytes.

A second event was B. McClintock's establishment of transposable elements in maize, as components of what she called the *Ac-Ds* system (1950). Who could think at that time of any chromosome region moving from chromosome to chromosome in an ordered fashion? Genes were considered to be stable structures that could mutate but not jump around. Although B. McClintock was much respected by her colleagues, due to her previous contributions to maize cytogenetics, few understood the significance of her results and her work was drowned in silence. One had to wait for a quarter of a century, to find a similar phenomenon in bacteria, that could be studied at the molecular level, to finally realize the significance of the *Ac-Ds* system. Results are not accepted in science, even if they are obtained by outstanding colleagues, and even if they are demonstrated with full accuracy. They are only accepted, and incorporated, when they fit into the general trend of thought or agree with the dominating intellectual fashion that prevails at a given time.

A third result was elucidating in this respect. In 1950 E. Chargaff was able to show, by chemical analysis of the bases of DNA, that adenine and thymine are present in equimolar amounts and that guanine and cytosine are also found in equimolar amounts. This regularity in the composition of DNA was referred to as Chargaff's rule. The discovery of the equivalence of the two groups of bases was of the utmost importance in the construction of the model of DNA that was proposed three years later by J.D. Watson and F.H.C. Crick.

The centromere was considered the most important region of the chromosome due to two major properties: 1) It was the region of active mobility on the spindle that allowed the chromosomes to move to the poles. 2) It was the segment that controlled the distribution of several chromosome properties such as chiasmata and crossing-over. However, the centromere remained an enigmatic region since it appeared in the microscope as a structureless "gap". The analysis of pachytene chromosomes in rye revealed a complex structure that could be divided into three regions. These built a reverse repeat consisting of chromomeres and fibrils. In the following year it was shown that this structure contained DNA. These findings established that the centromere did not differ from the chromosome arms either in structure or chemical composition (A. Lima-de-Faria 1949, 1950). This centromere pattern turned out to be of general occurrence and made it possible to explain the occurrence of iso-chromosomes (with symmetrical arms and a symmetrical centromere structure) which are found regularly in nature.

Fifth Period 1950–1970

MODEL OF DNA – WATSON AND CRICK – 1953
ELECTRON MICROSCOPY REVEALS THE ULTRASTRUCTURE OF CELL ORGANELLES
HIGH RESOLUTION FIGURES OF MITOCHONDRIA – PALADE – 1952
ELECTRON MICROGRAPHS OF CHLOROPLASTS – SJÖSTRAND – 1953
ENDOPLASMIC RETICULUM – PORTER – 1953
GOLGI APPARATUS – DALTON AND FELIX – 1954
LYSOSOMES DISCOVERED – DE DUVE – 1955
HUMAN CHROMOSOMES = 46 – TJIO AND LEVAN – 1956
PHOSPHORUS 32 – DNA SYNTHESIS OCCURRING AT ONE PERIOD OF INTERPHASE –
HOWARD AND PELC – 1953
TRITIUM – SEMI-CONSERVATIVE REPLICATION IN CHROMOSOMES –
TAYLOR ET AL. – 1957
TRITIUM – HETEROCHROMATIN REPLICATING LATER THAN EUCHROMATIN –
LIMA-DE-FARIA – 1959
GENETIC CODE – OCHOA, KORNBERG, NIRENBERG 1955 TO 1961
MESSENGER RNA – JACOB AND MONOD – 1961
LOCATION OF RIBOSOMAL RNA GENES – RITOSSA AND SPIEGELMAN – 1965
SYNTHESIS IN VITRO OF SELF-PROPAGATING RNA – SPIEGELMAN 1965 TO 1967
REPETITIVE DNA – BRITTEN AND KOHNE – 1968

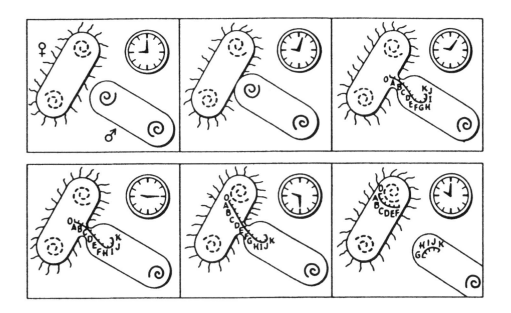

JACOB and WOLLMAN 1961 -
Schematic representation of the conjugation process between Hfr and F⁻ bacteria. Conjugation begins shortly (upper right) after the two bacteria come into contact, and is soon followed by the linear transfer of genetic material. The capital letters represent the location of various genes along the donor chromosomes.

Radioisotopes and electron microscopy became a most fruitful combination
Molecular biology received its main impulse from disciplines outside genetics

We tend to think that molecular biology arose mainly from biological disciplines such as genetics but that is far from true. Physics, crystallography, biochemistry and microbiology were the main sources of the new discipline.

Niels Bohr, as early as 1936, made his Institute of Physics at Copenhagen a research center, where one of the main themes was the study of the possible contribution of physics to the explanation of biological phenomena. M. Delbruck was the disciple of N. Bohr who carried this message further in his work in biology and in his book "A Physicist Looks at Biology" (1949). Another physicist, E. Schrödinger, in "What is Life" (1944), was the first to define the chromosome as "an aperiodic crystal". G. Gamow, also a leading physicist, involved in the building of the atom bomb, helped to predict the sequence of amino acids in the structure of proteins.

In biochemistry the contribution of W.T. Astbury was outstanding. He was the first to produce a model of DNA with the bases and the sugars stacked on top of each other (1945). Crystallography had an equally significant impact. J.D. Bernal contributed early with X-ray photographs of proteins (1934) and M.H.F. Wilkins and his collaborators (1953) delivered the X-ray analysis of DNA which allowed conceiving the spiral shape of this macromolecule. From microbiology came the pioneer contributions of A. Lwoff, F. Jacob, J. Monod and of S. Spiegelman.

Five years after the end of the Second World War, international relations among scientists were becoming reestablished. Science flourished again, ironically, in part thanks to the "Cold War" that followed the ceasing of the hostilities. To counteract Soviet expansion the Marshall plan was created to support the economic development of a partly destroyed Europe, and the Rockefeller Foundation contributed with grants to develop scientific research in Western Europe. Parallel to this effort was the creation of Research Councils in most of these countries, and in Germany the Kaiser-Wilhelm Institutes were transformed into Max-Planck "Abteilung". Technical and scientific development was given the highest priority by most governments. Atom bomb tests had become routine events and they were painful evidence of the new technological era that had to be mastered.

One tends to think of the importance of radioisotopes mainly in three ways: 1) Their significance in physics for the understanding of the atom structure. 2) Their importance in medicine as an agent of cancer therapy. 3) Their general use for peace and for war purposes in the form of nuclear reactors and atom bombs.

However, radioisotopes had a fourth role in science, that has not always been fully realized. Without them present cell biology would not have emerged. The knowledge of the molecular functions of the chromosome and of the cell, with all the consequences that it has had for the development of biology, medicine and agriculture, could hardly have been obtained without tagging molecules with specific radioisotopes. These allowed locating atoms with precision in cell structures, to follow their transit between the different organelles and finding out at what time they were incorporated into a given molecular edifice.

By 1942 R. Schoenheimer described the turnover of organic compounds in cells by using radioisotopes, but the field did not expand until phosphorus 32 and tritium (H^3) were introduced in cell biology. These radioisotopes emitted weak beta particles which travelled on the average only a few microns. This was a pre-condition for obtaining the desired resolution if one was going to get information about the function of minute cell organelles or macromolecules such as nucleic acids or proteins.

At the same time techniques employed in electron microscopy became refined. 1) More powerful and accurate microscopes were built. 2) Glass knives that allowed the production of thinner sections were introduced in 1950 by H. Latta and J.F. Hanmann. 3) The ultramicrotome became commercialized in 1953. 4) The diamond knife, which endured cutting tissues for a longer period, was developed by H. Fernandez-Moran (1956).

As a consequence of these improvements new results started to show up from many laboratories.

The period was heralded by G.E. Palade's first high resolution picture of mitochondria (1952). It was followed in the next year by the publication of electron micrographs of chloroplasts by J.B. Finean, F.S. Sjöstrand and E. Steinmann. But the major discovery was that of the endoplasmic reticulum by K.R. Porter (1953), that could now be seen clearly in electron micrographs. By 1954 the Golgi apparatus ceased to be an "artefact" and became a well defined organelle, as revealed by its ultrastructure (A.J. Dalton and M.D. Felix 1954). One year later C. de Duve and collaborators described a new organelle and coined the term *lysosome*.

The inner structure of the cell organelles had now become well defined and this brought its molecular architecture within reach. The moment seemed propitious to start using radioisotopes.

Autoradiography was a technique in which molecules labeled with radioactive elements were detected by the application of a photographic emulsion or a film (1953) to a tissue, or a cell squash. A. Howard and S.R. Pelc, using the new technique, demonstrated that there were periods, at interphase, during which DNA synthesis did not occur (1953).

But the commercialization of molecules labeled with tritium, which allowed a resolution of 1 to 2 microns, opened the way to studies at the chromosome level.

J.H. Taylor, P.S. Woods and W.L. Hughes (1957) used tritiated thymidine combined with autoradiography. By using *Vicia faba* chromosomes they found the semi-conservative distribution of label during chromosome replication. This work was soon followed (1958) by that of M. Meselson and F.W. Stahl, who by means of density gradient equilibrium ultracentrifugation demonstrated the semi-conservative distribution of labeled DNA in *E. coli*.

Heterochromatin continued to be an elusive component of the chromosome both in what concerned its gene content as well as its function. Sex chromosomes in certain insects had been known, for a long time, to be clear-cut cases of heterochromatic structures. By combining autoradiography with tritium labeled thymidine, A. Lima-de-Faria (1959), discovered that the heterochromatic sex chromosome of *Melanoplus* synthesized DNA later than the rest of the chromosome complement. This turned out to be a general phenomenon. Heterochromatic regions of plant, animal and human chromosomes were later found to also be late replicating.

In the same year M. Chevremont and collaborators, using the autoradiographic method showed that DNA was present in mitochondria (1959). By employing electron microscopy H. Ris and W. Plaut confirmed that chloroplasts contained DNA (1962). As early as 1951 Y. Chiba, by applying the Feulgen test, had demonstrated the existence of DNA in chloroplasts. But due to the small size of these organelles, this result was generally considered "an artefact". At that time no one was willing to accept that DNA existed outside the nucleus. But now suddenly DNA was present in two well defined cytoplasmic structures, which contained their own chromosomes and which were independent of those present inside the nuclear envelope.

This was a period so rich in discoveries, that took place in such a rapid succession, that it becomes necessary to sort them out in groups.

The most innovative proposal was the DNA model of J.D. Watson and F.H.C. Crick according to which the DNA molecule consisted of two strands bound by hydrogen bonds between their bases. This model was based primarily on the earlier contributions of W.T. Astbury, E. Chargaff and M.H.F. Wilkins. The combination of the biochemical and crystallographic data available, allowed understanding the replication of DNA by means of a simple mechanism. This finding (1953) had an enormous impact on future molecular work.

Soon after that, M.B. Hoagland obtained protein synthesis in cell-free systems (1955) and M. Grunberg-Manago and S. Ochoa isolated an enzyme participating in the synthesis of a nucleic acid – polynucleotide phosphorylase (1955). S. Ochoa and collaborators as well as A. Kornberg and associates obtained the *in vitro* enzymatic synthesis of ribonucleotides and deoxyribonucleotides (1956). In the same year G.E. Palade and P. Siekevitz isolated ribosomes, another crucial step in the discovery of the complex process of protein synthesis. Soon after, F.H. Crick predicted the existence of an adaptor molecule called transfer RNA (1958) and P.C. Zamecnik and collaborators identified transfer RNA molecules associated with amino-acids (1958). A key result sprang from the work of K. McQuillen, R.B. Roberts and R.J. Britten who showed that the ribosomes of *E. coli* were the sites of protein synthesis (1959). This work was followed by P. Siekevitz and G.E. Palade's evidence on the synthesis of proteins on ribosomes (1960). The existence of messenger RNA was predicted by F. Jacob and J. Monod (1961). The presence of this new molecule was soon confirmed by two independent groups led by S. Brenner and by F. Gros (1961). In the same year B.D. Hall and S. Spiegelman developed a technique which allowed isolating and characterizing messenger RNA (1961). At the same time F. Jacob and E.L. Wolman (1961) visualized the conjugation process in bacteria and the transfer of donor chromosomes.

F.H. Crick and colleagues predicted that the genetic code was divided into sets of three bases (1961) and M.W. Nirenberg and J.H. Mathaei broke the code by studying protein synthesis in cell-free systems (1961). They made the observation that when the synthetic polymer poly(U) was added to the system with mixtures of 20 amino acids, only one amino acid in each mixture being radioactive, the only amino acid to be incorporated was phenylalanine and the product was polyphenylalanine. The RNA code of this amino acid was thus found.

During the years 1963 and 1964 a series of findings led to a comprehensive picture of the complex mechanism of protein synthesis. R. Rosset and R. Monier isolated 5SRNA and T. Okamoto and M. Takanami discovered that messenger RNA attached to the small subunit of ribosomes. H. Noll and collaborators showed that protein synthesis occurred by a "tape" mechanism and the colinearity of gene and protein product, was established in the virus T4 and in *E. coli*. Finally W. Gilbert discovered that the transfer RNAs and nascent proteins appeared to be attached to the large subunit of ribosomes.

The interest then shifted to the *in vitro* synthesis

of RNAs. In 1965 S. Spiegelman and collaborators demonstrated the *in vitro* synthesis of self-propagating infectious RNA using a purified enzyme. This work carried out on bacteriophage Qbeta was followed by an *in vitro* evolution experiment that generated a self-duplicating molecule (1967). In the same year H.G. Khorana used di- and trinucleotides to elucidate the mechanism of the genetic code.

Equally important, as the J.D. Watson and F.H.C. Crick model, was another model proposed during this period. F. Jacob and J. Monod developed the concept of the operon in a paper published in 1961. This led to the distinction between structural genes (responsible for protein synthesis) and operator genes, as well as the introduction of such concepts as promoter, repressor, inducer and the ordered synthesis of messenger RNA. By 1969 the lac operon from *E. coli* had been isolated, in pure form, by J.R. Beckwith and associates, confirming the correctness of the original model.

A series of technological advances in chromosome research also occurred at this time.

The chromosome number of most mammalian species was not known as late as 1956. Several lines of research converged leading to the improvement of this situation. Pretreatments of cells led to better squashing and spreading of the chromosomes. Cancer cells were found to have variable numbers of chromosomes and as a consequence there was a demand for an exact analysis of chromosome morphology and chromosome number in tumours. This led J.H. Tjio and A. Levan to produce clear preparations of human chromosomes that allowed establishing the correct number of the karyotype (1956). One difficulty remained, biopsies of human organs usually gave bad chromosome spreads. This obstacle was removed when P. Nowell found that phytohemagglutinin induced the division of human leukocytes from peripheral blood, a technique that allowed the easy study of chromosomes from any individual (1960).

The medical consequences of this new technology were immediate. Diseases were now found to be correlated with specific chromosome aberrations. J. Lejeune and collaborators discovered that the Down's syndrome in humans was accompanied by a trisomy of a small chromosome (1959). In the same year C.E. Ford and associates coupled the Turner's syndrome to an X0 condition. This was followed by relating the Klinefelter's syndrome to an XXY chromosomal set (1959). In 1961 M.F. Lyon and L.B. Russel, working in different laboratories, proposed the inactivation of one of the two normal X chromosomes of the mammalian female. Since then human cytogenetics expanded exponentially becoming one of the basic disciplines of medicine.

The work on chromosomes was accompanied by a parallel development in nucleic acid research.

B.D. Hall and S. Spiegelman produced molecules consisting of one strand of RNA and one strand of DNA. These hybrid molecules allowed establishing the complementarity between messenger RNA and specific DNA sequences, which led to the isolation of messenger RNAs (1961). A year earlier, P. Doty and associates, had demonstrated that complementary strands of DNA could be separated and recombined. These two results contributed to build the basis of modern DNA technology. It then became urgent to devise a method to identify each single base. To simplify matters one started with RNA and with a small one. As a result R.W. Holley and collaborators obtained the complete base sequence of alanine transfer RNA extracted from yeast cells (1965).

Until then, most genes and DNA sequences were supposed to occur in only one or a few copies. This situation was reversed by a finding made by R.J. Britten and D.E. Kohne. By separating and reassociating the DNA strands they demonstrated the occurrence of repetitive DNA in higher organisms (1968).

New methods of gene mapping also became available.

B.C. Westmoreland, W. Szybalski and H. Ris mapped genes in lambda viruses with the help of electron microscopy (1969). A new method, which allowed localizing human genes, resulted from the formation of cell hybrids between different species. Somatic cell genetics became an independent line of research. Thymidine kinase was localized by M.C. Weiss and H. Green, using this method in 1967.

The biochemical and genetic research carried out in viruses and bacteria had been so fruitful in analyzing their molecular mechanisms, that it invited an attempt to use the same methodology at the level of the higher organisms. The techniques that had worked well with prokaryotes should now be amenable to probing the cells of animals and humans. The molecular analysis of the chromosomes of higher organisms became the next goal.

Ribosomal RNA was the most abundant species of RNA to be found in animal cells, and as such could be easily extracted in large amounts and in pure form. One thus started to search for the location of the genes that would be responsible for its production.

The synthesis of 18S and 28S ribosomal RNA was soon found to be located in the nucleolus organizer region of the chromosomes. This became possible thanks to a deficiency for this region in the toad *Xenopus* (D.D. Brown and J.B. Gurdon 1964). This finding was followed by the localization of multiple copies of these genes in the X and Y chromosomes of *Drosophila* by F.M. Ritossa and S. Spiegelman (1965). This work culminated in two new findings: the establishment of gene amplification and the discovery of hybridization *in situ*, a technique that to this day, has produced invaluable information about gene location and chromosome organization.

Gene amplification was demonstrated in insects

and *Xenopus* between 1968 and 1969 by several independent research groups that included A. Lima-de-Faria, D.D. Brown, M. Birnstiel, and J. Gall.

A similar situation occurred concerning the hybridization *in situ*. Since the DNA-RNA hybridization had become possible by adding RNA to filters that contained different DNA fractions; it became natural to contemplate the possibility of hybridizing the RNA directly to the DNA of chromosomes present in fixed cell preparations. As with the rediscovery of Mendel's "laws" the same result was obtained by three different research groups working in three different countries in 1969 and 1970. These were H. John, M.L. Birnstiel and K.W. Jones (Scotland), J.G. Gall and M.L. Pardue (U.S.A.) and M. Buongiorno-Nardelli and F. Amaldi (Italy).

This technique was soon combined with electron microscopy allowing *in situ* analysis of DNA molecules in electron micrographs. It became possible to go even one step further and to visualize, in the electron microscope, nascent RNA molecules which were formed along the DNA molecule building what was called a "christmas tree" (O.L. Miller and B.R. Beatty 1969).

Another area of utmost importance, whose significance was not fully realized at that time, was the self-assembly of viruses from their isolated constituents. H. Fraenkel-Conrat (1962) isolated separately the RNA and the protein of tobacco mosaic virus. When appropriate pH conditions were supplied these two macromolecules self-assembled and the virus particles formed were fully infectious and indistinguishable from the original virus. In future decades it became possible to produce by self-assembly such cell organelles as nucleosomes, ribosomes and cell membranes. This showed how atomic mechanisms, inherent to the cell molecules, determined the construction and organization of the chromosome and of the cell.

This is also the period in which the three-dimensional structures of proteins were starting to be revealed. J.C. Kendrew and associates obtained myoglobin at 2 Å resolution and M.F. Perutz and collaborators hemoglobin at 5.5 Å resolution (1960). C.C.F. Blake published the structure of lysozyme at 2 Å resolution in 1967.

In the meantime the functions of the nucleus and of the chromosome continued to be the focus of much attention. R. Briggs and T.J. King enucleated frog's eggs and introduced nuclei from blastula cells into these eggs. The transplanted nuclei survived and underwent chromosome changes (1952). Following this trend of thought 10 years later J.B. Gurdon obtained normal frog adults derived from an enucleated egg injected with a nucleus from an intestinal cell. This was taken to mean that the gene constitution and potency were identical in somatic and germ cell nuclei (1962). By 1967 L. Goldstein and D.M. Prescott carried out nuclear transplantations in *Amoeba* showing that proteins moved from the cytoplasm to the nucleus.

The individual chromosomes, and within them regions that could be recognized by their specific phenotype, attracted the attention of cytologists. W. Beermann discovered that certain "puffs" of polytene chromosomes were characteristic of certain tissues and postulated their relationship to gene activity (1952). As a result C. Pelling discovered the differential labelling of puffed regions, using H^3 uridine which incorporated into RNA, and which also disclosed gene activity (1959). C. Pavan (1959) analysed the sequence of band morphology and measured the increase of DNA content occurring during the puffing of specific bands in the salivary gland chromosomes of *Rhynchosciara*. F.M. Ritossa obtained the puffing of bands in *Drosophila* following heat shock treatments during which the production of specific proteins took place.

During the three ensuing decades one would reap the fruits resulting from the ground work laid down during this period.

Sixth Period 1970–1980

BANDING OF HUMAN CHROMOSOMES WITH QUINACRINE DYES –
CASPERSSON ET AL. – 1970
SYNTHESIS OF THE GENE FOR AN ALANINE TRANSFER RNA – KHORANA ET AL. – 1970
RNA-DEPENDENT DNA POLYMERASE IN ONCOGENIC RNA VIRUSES –
BALTIMORE AND TEMIN – 1970
RESTRICTION ENZYMES BECOME PHYSICAL TOOLS – DANA AND NATHANS – 1971
GENES RESPONSIBLE FOR BODY SEGMENTATION IN DROSOPHILA –
GARCIA-BELLIDO ET AL. – 1973
DNA SEQUENCING TECHNIQUE – SANGER AND COULSON – 1975
TETRACARCINOMA CELLS REMAIN TOTIPOTENT – MINTZ AND ILLMENSEE – 1975
DNA SEQUENCES TRANSFERRED TO NITROCELLULOSE FILTERS – SOUTHERN – 1975
FUSION OF HUMAN AND PLANT CELLS – DUDITS ET AL. – 1976
GENES SEPARATED IN EMBRYONIC CELLS BECOME ADJACENT LATER IN
DEVELOPMENT – HOZUMI AND TONEGAWA – 1976
GENE LIBRARIES TO ISOLATE GENES – MANIATIS ET AL. – 1978
TRANSGENIC MICE, DNA INJECTED INTO PRONUCLEUS OF A FERTILIZED EGG –
GORDON ET AL. – 1980

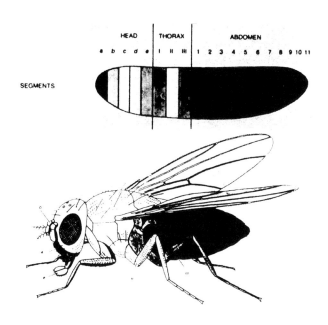

GARCIA-BELLIDO et al. 1979 -
Compartments in the development of the fruit fly *D. melanogaster* are indicated by means of schematic diagrams of the embryo and the adult fly. The embryo, a peripheral epithelium of some 5000 cells descended from a single nucleus in the egg, becomes divided into segments, each destined to give rise to a segment of the adult.

The mechanisms of cancer and of development were sought at the DNA level

Biotechnology emerged as a new field as genetics created its own weapons

Although this was a period of only ten years, it led to the harvest of many new results. Also some basic new techniques were created in this short time.

These findings can be grouped into several areas.

1. Improvement in the resolution of chromosome segments and gene location

T. Caspersson had remained silent for a long time after his contribution in the 1930s and 1940s. Working in collaboration with L. Zech and C. Johansson he used quinacrine dyes which resulted in differential banding along human chromosomes (1970). These bands turned out to be characteristic of specific chromosomes, which allowed distinguishing human chromosomes that could not be identified before with accuracy. This was the starting signal for a whole field of research, in which other investigators used different chromosome treatments to elicit the formation of specific bands, not only in human chromosomes, but in those of other species as well, including plants.

The location of genes in human chromosomes was improved by the use of the banding method leading in turn to the location of specific genes connected with cancer. H.J. Evans and collaborators (1971) identified with accuracy the 22 pairs of human autosomes and established that the abnormal Philadelphia chromosome (connected with cancer) was an aberrant chromosome 22. This was followed (1972) by the BUDR labelling technique (A.F. Zakharov and N.A. Egolina) that produced harlequin chromosomes.

2. Bringing cancer to the DNA level

D. Baltimore and H.M. Temin discovered an RNA-dependent DNA polymerase in oncogenic RNA viruses in 1970. By 1974 I. Zaenen et al. recognized the tumor-inducing plasmid of the crown gall bacterium, which causes tumors in plants. In the same year B. Ames improved a method to detect carcinogenic compounds. L.H. Wang et al. (1975) located the RNA sequence responsible for the oncogenic activity of the Rous sarcoma virus. Using the same virus M.S. Collet and R.L. Erickson (1978) found that the product of the *src* gene was a protein kinase. B. Mintz and K. Illmensee using cell fusion and foster mothers in mice, were able to demonstrate that the nuclei of tetracarcinoma cells remained totipotent after many generations in which they had behaved as malignant cancers (1975). By 1980 L. Olson and H.S. Kaplan produced human hybridomas that manufactured a specific antibody in laboratory cultures.

3. The emergence of biotechnology as an independent area of research

A starting point in this direction was the introduction of DNA into *E. coli* cells by M. Mandel and A. Higa (1970). In the same year H.G. Khorana and colleagues, using yeast cells, synthesized the gene for an alanine transfer RNA. C.R. Merril et al. used cultured fibroblasts from a patient suffering from galactosemia.

Subsequently a lambda phage, carrying the galactose operon, was transduced into the fibroblasts, which acquired the ability to synthesize the necessary transferase and survived in culture (1971). Restriction enzymes became laboratory tools when K. Dana and D. Nathans used them to cleave DNA of simian virus 40. The fragments obtained could be ordered after their physical length (1971). Such DNA fragments could soon be rejoined using the method of P. Lobban and A.D. Kaiser (1972). After that R. Silber et al. discovered RNA ligase (1972). The technique evolved rapidly and in the same year D.A. Jackson et al. joined the DNAs of the viruses SV40 and lambda. J. Mertz and R.W. Davis described the cleavage of DNA by the restriction endonuclease RI which generated cohesive ends (1972). S.N. Cohen et al. (1973) combined *in vitro* different plasmids and W. Fiers et al. sequenced a gene that coded for a protein in the phage MS2 (1973).

K.M. Murray and N.E. Murray made of lambda a cloning vehicle by using restriction enzyme technology (1974). This was followed by the work of M. Grunstein and D.S. Hogness who developed a technique that allowed isolating cloned DNAs containing specific genes (1975), and in the same year E.M. Southern invented the procedure permitting the transfer of DNA sequences from agarose gels to nitrocellulose filters. Following hybridization with radioactive RNA, the DNA-RNA hybrids were detected by autoradiography. This technique had a great impact since it allowed, among other things, the comparison of the DNAs from different organisms.

W.Y. Kan et al. used recombinant DNA technology to develop a prenatal test (1976). By 1977 the commercial production of the first artificial human protein, somatostatin, was achieved by synthesizing this gene and expressing it in *E. coli* (K. Itakura and collaborators 1977). In the same year W. Gilbert obtained insulin and interferon that were synthesized by bacteria.

The method developed by E.M. Southern for DNA analysis stimulated the development of a comparable method for RNAs, in which these molecules were hybridized with known radioactive DNA sequences. This technique was called "Northern blotting" since it was the reverse process of the Southern method (J.C. Alwine et al. 1977). Subsequently T. Maniatis and collaborators prepared gene libraries to isolate genes (1978).

Biotechnology had become a reality, and the mammalian, as well as the human egg, were from now on objects of genetic experimentation.

4. *Somatic cell fusion and the problem of cell totipotency*

It had been known for a long time that most plant cells are totipotent, *i.e.* from a single piece of a leaf it is possible to obtain a complete plant that produces flowers and fruits, as is, for instance, the case in *Begonia*.

This totipotency is absent in most animal and human cells. If this property could be transferred from plant to human cells it would become possible to obtain embryos from a series of human tissues.

To date this transfer remains elusive because we continue to lack knowledge of the genetic properties of several cytoplasmic organelles. In the meantime new techniques developed, for the fusion of plant cells, and for the removal of their cell walls. A first attempt to elucidate this problem, which is also of significance for the understanding of the origin of cancer, was carried out when human cells were fused with carrot protoplasts by using polyethylene glycol as the fusing agent (D. Dudits et al. 1976). Nuclear fusion and cell wall formation occurred in the plant-human cell hybrids. A similar result was obtained independently by C.W. Jones et al. (1976).

In a second series of experiments A. Lima-de-Faria et al. (1977) used sendai virus to fuse *Haplopappus* protoplasts (plant cells from which the cell wall had been removed by enzymes) with human cells. The fused human and plant nuclei, as well as their cytoplasms, could be distinguished by means of tritium labeling and staining procedures. In later experiments (A. Lima-de-Faria et al. 1984) the human chromosomes were found to be dividing in the plant-human cytoplasm.

Results that contributed to open this field were the work on animal cells of R.T. Johnson and P.N. Rao who fused mitotically active cells with cells at interphase. This fusion resulted in the premature condensation of the chromosomes (1970) indicating an interaction between the fused nuclei. Soon after (1973) P. Debergh and C. Nitsch obtained fertile tomato plants which had originated from single microspores growing in culture, a result that confirmed the totipotency of plant cells.

5. *The discovery of several DNA sequencing techniques opened the way for accurate genetic comparisons*

Two techniques for sequencing DNA were discovered during this period. They revolutionized the field of molecular biology since every base found along a DNA molecule could now be determined. F. Sanger and A.R. Coulson developed the first method by primed synthesis with DNA polymerase (1975). Two years later A.M. Maxam and W. Gilbert discovered a second method which allowed to sequence DNA chemically (1977). In this year F. Sanger and associates obtained the complete nucleotide sequence of the bacterioiphage phiX174 (1977).

6. *Drosophila again became the organism of choice – It allowed unravelling the mechanism responsible for development*

Development of the fertilized egg into an adult organism continued to be one of the main black boxes of genetics. There was a deep ignorance of the genetic mechanisms responsible for cell location and cell differentiation during embryo formation. Since so much knowledge of genes was available in *Drosophila*, this organism again became the experimental animal of choice. The effort paid well by furnishing startling results in a relatively short time. A. Garcia-Bellido and collaborators initiated the study of development by analysing genes responsible for body segmentation and especially the discs involved in wing formation (1973). E.B. Lewis soon found that the genes of the *bithorax* complex had related functions which were coupled to *Drosophila* segmentation (1978). Subsequently C. Nüsslein-Volhard and E. Wieschaus isolated mutations responsible for body segmentation in this fly (1980). The field was now open.

7. *Split genes and transposons were discovered. These were results that later had unexpected consequences for the understanding of evolution*

The period started with the localization of genes at specific chromosome sites. D.E. Wimber and D.M. Steffensen localized the 5S ribosomal RNA genes on the right arm of chromosome 2 of *D. melanogaster* (1970) and the histone genes were located on chromosome 2 of the same species by M.L. Pardue et al. (1972). In the same year L.P. Suzuki and D.D. Brown characterized the messenger RNA of the fibroin gene in the silk worm (1972). Y. Hotta and H. Stern found a later DNA synthesis at meiosis that could be responsible for the process of crossing-over (1971).

At that time it was maintained by several workers that the DNA of chromosomes was interrupted by protein segments. R. Kavenoff and B.H. Zimm measuring the molecular weights of the DNAs isolated from *Drosophila* species, asserted that each chromosome consisted of a single DNA molecule that traversed the centromere (1973). The molecular organization of the chromosome was further elucidated when R.B. Kornberg suggested that the histones and the DNA of chromosomes were associated in units comprising 200 base pairs of DNA. These were later called nucleosomes and were visualized in the electron microscope (1974).

The term transposon, used for mobile chromosome segments, was coined by R.W. Hedges and A.E. Jacob (1974) following their studies in *E. coli*. This result revived interest in the work of B. McClintock who had demonstrated the existence of mobile elements in maize as early as the 1950s. N. Hozumi and S. Tonegawa produced the first evidence that genes physically separated in embryonic cells of mice, became adjacent in plasmacytomas. During maturation of B lymphocytes the genes for the variable and constant regions of an immunoglobulin chain moved close together during differentiation (1976).

The following year saw the discovery of the split gene. This was achieved by three different groups. They worked with: 1) adenovirus (R.J. Roberts, P.A. Sharp); 2) rabbit beta-globin genes (A. Jeffreys and R.A. Flavell); and 3) the chicken ovalbumin gene (R. Chambon and collaborators) (1977). This discovery disclosed that segments of the messenger RNA were discarded and did not participate in the RNA that coded for the protein. In the following year W. Gilbert coined the terms intron and exon.

Telomere regions were studied for the first time at the molecular level and were found to be characterized by DNA sequences repeated tandemly in *Tetrahymena* by E.H. Blackburn and J.G. Gall (1978).

8. Order started to become evident in the emergence of mutations

The early work with X-rays led to the idea that mutations were random events. The radiation had such a drastic effect on the chromosome that it demolished it, causing all kinds of rearrangements. Later, when chromosomes were treated with chemicals, it became evident that certain types of rearrangements were more common than others. X-rays and other types of radiation played havoc on the chromosome, but chemicals turned out to be more specific in their activity pointing to an ordered process in the production of mutations.

More refined molecular methods revealed that mutations were mainly non-random events. C. Coulondre et al. showed that in *E. coli* mutational hot spots were present in DNA sequences that contained modified bases (1978), and C.A. Hutchison et al. revealed that specific mutations could be introduced at proper sites in the DNA molecule (1978). However, a result with particular evolutionary significance was obtained by R.T. Schimke et al. who found that mouse cells in culture when exposed to drugs could amplify the genes that led the cells to become resistant to the toxic substance (1978).

9. Chloroplast and mitochondrial DNAs entered the test tube

R. Sager and Z. Ramanis localized in the chloroplast of *Chlamydomonas* a group of eight genes. This was the first map of genes not situated in the nucleus (1970). The hypothesis had been advanced for several years that chloroplasts had entered the eukaryotic cell as free living organisms and were descendants of blue-green algae. This idea received support from the hybridization of ribosomal RNAs extracted from living cyanobacteria (blue-green algae) with the chloroplast DNA of the eukaryote *Euglena gracilis* (G.H. Pigott and N.G. Carr 1972).

The mitochondria also became the source of active investigation. B. Dujon et al. discovered that DNA recombination occurred in the DNA of yeast mitochondria (1974) and maternal inheritance of mitochondrial DNA, in horse-donkey hybrids, was described by C.A. Hutchison et al. (1974). More startling was the finding that the genetic code, which had been found to be the same in all organisms so far investigated, turned out to be different in the DNA of human mitochondria (B.G. Barrell et al. 1979).

Seventh Period 1980–1990

HUMAN HYBRIDOMAS THAT MANUFACTURE A PURE ANTIBODY –
OLSSON AND KAPLAN – 1980
RESTRICTION LENGTH POLYMORPHISM TO CONSTRUCT LINKAGE HUMAN MAPS –
BOTSTEIN ET AL. – 1980
ISOLATION OF SEGMENTATION GENES IN DROSOPHILA –
NÜSSLEIN-VOLHARD AND WIESCHAUS – 1980
TRANSFER OF PHASEOLIN GENE OF BEANS, VIA PLASMID, TO SUNFLOWER –
KEMP AND HALL – 1981
COMPLETE SEQUENCING OF MITOCHONDRIAL GENOME IN HUMANS –
SANGER ET AL. – 1981
MOLECULAR CONTROL OF MEMORY IN APLYSIA, MEMORY GENES –
KANDEL AND SCHWARTZ – 1982
LOCALIZATION OF TRANSCRIPTS OF HOMEOTIC GENES IN DROSOPHILA –
HAFEN ET AL. – 1983
DNA FINGERPRINT TECHNIQUE – JEFFRIES ET AL. – 1985
POLYMERASE CHAIN REACTION – SAIKI ET AL. – 1985
RNA EDITING IN TRYPANOSOMES – BENNE ET AL. – 1986
INTRODUCTION OF A HUMAN GENE INTO THE MOUSE – KUEHN ET AL. – 1987

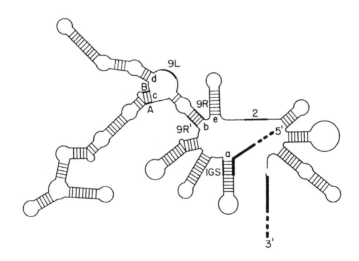

CECH and BASS 1986 -
Mechanism of splicing of transcripts containing group I intron in the pre-ribosomal RNA of *Tetrahymena*. The RNA splicing occurs in vitro in the absence of proteins.

Neurobiology reached the molecular level
Artificial chromosomes and gene therapy became a reality

It was Holger Hyden, a neurophysiologist, who in the early 1960s, at the University of Gothenburg in Sweden, showed that neural processes were amenable to molecular analysis, by studying the amount of RNA in rat cells, before and after training. The technique at that time was in its infancy. It did not allow going further. One had to wait 20 years, as molecular biology became supplied with a series of new tools, to see the arrival of neurobiology as an independent field.

1. The first steps in neurobiology

Among the eukaryotes, yeast looked too simple, but on the other hand *Drosophila* seemed too complex. One searched for a multicellular eukaryote that contained few organs and few cells. The nematode *Caenorhabditis elegans* became the choice. Since then hundreds of papers have been published on the genetics of this worm. One started to look for mutants connected with its behaviour. M. Chalfie and J. Sulston described in 1981 five genes that were related to the function of six sensory neurons in *C. elegans*. Another animal was also investigated. The molluscs are a large group of invertebrates which comprise species that are functionally and morphologically most disparate. This group also became attractive as a source of studies in animal behaviour. *Aplysia* became the choice mainly due to its giant neurons. E.R. Kandel and J.G. Schwartz studied a gill reflex in *Aplysia* to investigate memory at the molecular level. They concluded that there were memory genes and that their function was related to the activation of cyclic AMP (1982). Two years later T.A. Bargiello and M.W. Young cloned and sequenced a gene that controlled the biological clock and W. Herr et al. reported the existence of a binding domain encoded by homeotic genes that were expressed only in the nervous system (1988).

2. Genes that suppressed the growth of cancer cells

Oncogenes became the target of molecular analysis. E.P. Reddy et al. found that the substitution of a single base in the oncogene, responsible for human bladder carcinoma, resulted in the activation of this gene (1982). M. Rassoulzadegan et al. (1983) discovered that a portion of the protein coded by a viral gene immortalized cultured cells from rat embryos. H.M. Ellis and H.R. Horvitz isolated genes from *C. elegans* that led to the programmed death of cell groups (1986). S.J. Baker et al. suppressed the growth of human cancer cells by introducing specific genes into the chromosome complement (1990). At present more than 200 suppressor genes have been located on human chromosomes.

3. Human genes were studied with the aim of correcting disease

A general effort was made to develop human gene therapy. D. Botstein et al. constructed genetic linkage maps of the human genome by employing restriction enzymes to cleave DNA (1980). J. Gitschier et al. cloned the human gene of the antihemophilic factor (1984). Suddenly genetics became an instrument of law enforcement (1985). A.J. Jeffries et al. established the so-called "DNA finger print" technique. Simple tandem repetitive sequences were found to be scattered throughout the human genome which shared a common core sequence. Significant was that the DNA pattern obtained on each gel was characteristic of a given individual. This allowed identifying indivduals involved in crimes on the basis of very small DNA samples.

M.R. Kuehn et al. were able to transfer a human gene into mouse embryonic germ cells and to add them to mouse embryos, which became chimaeras. The result was the production of mice carrying human genes (1987). L.-C. Tsui et al. predicted the amino acid sequence of the protein coded by the gene for cystic fibrosis (1989), and F.D. Hong et al. found that the retinoblastoma gene contained 27 exons (1989).

The human glucocerebrosidase gene was found to mutate causing Gaucher's disease (M. Horowitz et al. 1989). These results led to the establishment of gene therapy in the following decade.

4. Mitochondria and chloroplast DNAs were sequenced

The entire mitochondrial genome was now sequenced in this human organelle by F. Sanger and collaborators (S. Anderson et al. 1981).

The complete nucleotide sequence and gene organization of the chloroplast were established in two unrelated species, *Marchantia polymorpha* (a liverwort) and tobacco plants (K. Ohyama et al. 1986 and K. Shinozake et al. 1986). In the same year M.-C. Shih et al. discovered that genes harbored in the ancestors of chloroplasts were transferred from these organelles into the chromosomes of the nucleus.

Leber's disease in humans was found to be due to a mutation in mitochondrial DNA (1988).

5. RNA editing

Artificial chromosomes became a reality when A.W. Murray and J.W. Szostak (1983) produced them for the first time in yeast. This was followed by the work of J.L. Harrington et al. who obtained human artificial chromosomes in 1997.

One of the most significant discoveries of this period was RNA editing. Its significance for the understanding of the evolution of the chromosome and of the cell has not yet been fully realized. RNA editing actually means that the RNA in a single stroke changed the message that it received from the DNA. Hence one does not need a mutation to produce a different protein, and more important, one does not need to change the DNA molecule to create a new function. Moreover one does not need a long period of time to create a new molecular edifice. This event modified radically what had before been considered one of the basic processes of evolution. The ability to evolve did not reside in DNA alone and did not necessarily reside in its capacity to mutate. RNA by itself, when its chemistry found it appropriate, changed the evolutionary scenario and did it instantaneously.

R. Benne et al. discovered the phenomenon of RNA editing in a trypanosome (protozoa) in 1986. This was a process in which the genetic information changed at the level of the RNA. The coding sequence of the RNA differed from the sequence of the DNA from which it was transcribed. The protein coded became different. In the mitochondria of trypanosomes, several bases are systematically added or deleted from the messenger RNA.

The same phenomenon also occurred in mammals in which different tissues, such as intestine and liver, produced different proteins derived from the same original RNA. The evolutionary change was caused by a simple codon change from CAA to UAA which affected termination and resulted in a much smaller protein being produced. Four years later (1990) it turned out that RNA editing was the result of RNA molecules, which guided the splicing process, functioning as molecular midwives (B. Blum et al.). Most important was that the RNA splicing occurred in vitro in the absence of proteins (T.R. Cech and B.L. Bass 1986).

6. PCR and other techniques

Biotechnology expanded into new areas. Transgenic mice were obtained by J.W. Gordon et al. (1980) by injecting a DNA segment into the pronucleus of a fertilized egg and M.R. Capecchi was able to transform mammalian cells in culture by injection of DNA with micropipettes (1980). Defective genes in mice could be substituted by normal ones following injection into fertilized eggs of these last genes. Important was the fact that the transgenic mice transmitted the normal condition to the offspring (F. Constantini et al. 1986). Plant genes were transferred among species belonging to different families. The phaseolin gene from beans was inserted into the genome of a sunflower using as a vector a plasmid of the bacterium *A. tumefaciens* (J.D. Kemp and T.H. Hall 1981). Another advancement in plant biotechnology occurred when it was demonstrated that the ability to fix nitrogen in R*hizobium* bacteria was due to plasmid-linked genes (G. Hombrecher et al. 1981). R. Mann et al. employed the Moloney murine leukemia virus as a vector in transfer experiments among mammalian cells (1983).

Agarose gel electrophoresis could only separate molecules of 50 kbp or less. D.C. Schwartz and C.R. Cantor introduced the pulse field gradient electrophoresis that could separate large DNA fragments of as much as 2,000 kbp (1984).

The polymerase chain reaction (PCR) was the technique that had the most revolutionizing impact on DNA technology (1985). R.K. Saiki et al. were able to amplify *in vitro* DNA fragments of the beta-globin gene. The number of amplified copies of a small DNA sequence could be as high as 1 billion. The technique demanded a control over undesirable DNA sequences that may occur during the procedure. Its results must be rigorously checked.

7. Techniques that improved detection of genes in chromosomes

Following the development of *in situ* hybridization that started in 1969 there was a search for successive improvements that allowed visualizing the location of genes with a higher degree of resolution, that is, that would make it possible to find the location of DNA sequences that occurred in only a few copies.

M.E. Harper and G.F. Saunders were able to map single-copy genes on human mitotic chromosomes (1981) and in the same year P.R. Langer et al. obtained coloured specific sites on chromosomes by synthesizing biotinylated DNA linked to streptavidin. This last molecule was attached to a colour generating agent. The technique was not only less time consuming than autoradiography but disposed with the undesirable use of radioisotopes. Moreover it improved the resolution (1981).

8. Development and the homeobox

As mentioned above, mutants that were involved in the body segmentation of *Drosophila* were isolated by C. Nüsslein-Volhard and E. Wieschaus (1980). As a consequence genes that controlled development also became known in the nematode *C. elegans* (I.S. Greenwald et al. 1983).

Gene clusters that were responsible for the formation of body segments in *Drosophila* were soon described at the molecular level by M.P. Scott and by W. Bender (1983). A new breakthrough came with the technique developed by W.J. Gehring and co-workers (1983) which allowed *in situ* hybridization of labeled DNA probes to RNA transcripts in frozen tissue sections. The homeotic genes were then allocated to different compartments of *Drosophila*.

One of the most significant results for the understanding of the universality of the mechanisms of development, as well as of evolution, sprang from the finding that the homeobox sequence from a fly (*Drosophila*) was present in a mammal such as the mouse (W. McGinnis et al. 1984). This view was reinforced by the discovery that regulatory genes contained conserved domains (M. Noll et al. 1986).

The impressive confirmation of this universality emerged when J. Malicki et al. transferred a homeobox gene from a mouse into the embryos of a fly (*Drosophila*), the result being that it produced a similar development pattern. This means that genes from species that have been separated by millions of years of evolution produced the same type of basic pattern and were interchangeable (1990).

9. *Ordered mutation – transposons*

Mutation appeared more and more ordered for every new experiment that was performed. Plants were able to become resistant to herbicides, just like bacteria that become resistant to antibiotics, or like mammalian cells that survive when treated with toxic substances. In every case the genome invented a solution that immediately counteracted the adverse situation imposed by the environment. In other words the cell refused to accept death, it simply mobilized its original construction circumventing and overcoming death, and simply could not wait passively for its destruction.

This was demonstrated in *Amaranthus*, (a weed) which was exposed to herbicides. Resistant strains emerged that modified a chloroplast gene that refused to bind the herbicide and allowed the plant to survive, circumventing death and selection (K.S. Steinbeck et al. 1981). J. Cairns et al. (1988) showed that the occurrence of mutations was dependent on chromosome organization. But it was by analysing the functions of transposons that the ordered processes of mutation became evident. There were found as many as 30 to 50 copies of transposable elements per genome in *Drosophila*. These integrated into the germ line and could function as vectors of other DNA sequences (G.M. Rubin and collaborators 1982). Subsequently M.K. Bhattacharyya et al. (1990) found a mutation in peas that resulted from the insertion of a transposon in a gene coding for an enzyme. This was another example of the role of a transposon in creating mutations.

10. *The revival of gradients and fields in embryonic development and chromosome organization*

Gradients of molecules were found to be present in fertilized eggs and embryos as early as the turn of the century. This was disclosed by staining methods that showed clearly the successive increase or decrease of a given substance. Moreover experiments that excised single cells (or cell groups) from embryos, led to the establishment of the concept of embryonic field. Regions of the embryo were found to develop differently depending on their position within the field. This finding was attributed to the distribution of the gradients. The rapid expansion of genetics with its concentration on gene localization and on the effects of structural rearrangements divorced genetics from embryology. Development could not be logically incorporated into the available knowledge of mutations. The result was that gradients and fields were rejected as sources of embryonic order. Such a point of view not only prevailed in development but extended to chromosomes. This attitude still dominates at present. However, the first blow came when the molecular level was reached and one found that gradients actually existed and that they were constituted by messenger RNAs, proteins and other molecules. Suddenly from being a "dirty word", as pointed out by P.A. Lawrence (1992) it became a highly fashionable term. Protein gradients were described by W. Driever and C. Nüsslein-Volhard along the antero-posterior axis of the embryo of *Drosophila* (1988), and messenger RNA was localized in the anterior region of the oocyte of the same species (P.M. Macdonald and G. Struhl 1988). The result is that fields are now being considered to explain development.

The same situation is occurring in the case of gradients and fields present in chromosomes. At the prophase of meiosis (pachytene) many plant and animal species exhibit a chromomere size gradient. The chromomeres are, in every investigated species, larger on both sides of the centromere and decrease in size towards the telomeres, building a gradient. Furthermore, like in eggs, the rate of decrease of the gradient builds a field, the position of genes being dependent on the length of the centromere-telomere region (A. Lima-de-Faria 1954, 1973, 1980). For a long time this phenomenon was difficult to interpret due to lack of molecular knowledge, but recent data support the chromosome field concept by showing that interactions between telomeres, centromeres and other DNA segments, located along the arms, affect their structure and function (H.A. Wichman et al. 1991, S. Henikoff 1992, P. Slijepcevic 1998, see page 176).

Eighth Period 1990–2001

FIRST EXAMPLE OF HUMAN GENE THERAPY, ENZYME DEFICIENCY DISEASE CORRECTED BY NORMAL GENE – ANDERSON – 1990
COMPLETE NUCLEOTIDE SEQUENCE OF CHROMOSOME III OF YEAST – OLIVER ET AL. – 1992
MEMORY GENES ISOLATED IN DROSOPHILA – TULLY ET AL. – 1994
COMPLETE NUCLEOTIDE SEQUENCE OF MYCOPLASMA – FRAZER ET AL. – 1995
ARTIFICIAL HUMAN CHROMOSOMES – HARRINGTON ET AL. – 1997
CLONING OF SHEEP DOLLY – WILMUT ET AL. – 1997
PROPOSED CLONING OF HUMANS – 1998
PROPOSED PRODUCTION OF HUMANS WITHOUT HEADS AS ORGAN DONORS, SPARE PARTS – 1998
SEQUENCING OF HUMAN CHROMOSOME 22 – DUNHAM ET AL. – 1999
SEQUENCING OF A FLOWERING PLANT GENOME – THE *ARABIDOPSIS* GENOME INITIATIVE – 2000
SEQUENCING OF THE HUMAN GENOME – THE INTERNATIONAL HUMAN GENOME SEQUENCING CONSORTIUM AND VENTER ET AL. – 2001

Organism	Year	Millions of bases sequenced	Total coverage (%)	Coverage of euchromatin (%)	Predicted number of genes	Number of genes per million bases sequenced
Saccharomyces cerevisiae	1996	12	93	100	5,800	483
Caenorhabditis elegans	1998	97	99	100	19,099	197
Drosophila melanogaster	2000	116	64	97	13,601	117
Arabidopsis thaliana	2000	115	92	100	25,498	221
Human chromosome 21	2000	34	75	100	225	7
Human chromosome 22	1999	34	70	97	545	16
Human genome rough draft (public sequence)	2001	2,693	84	90	31,780	12
Human genome rough draft (Celera sequence)	2001	2,654	83	88–93	39,114	15

BORK and COPLEY 2001 - Sequenced eukaryotic genomes. Total coverage uses an estimate of the total genome size and includes heterochromatin (condensed genomic areas that were originally characterized by staining techniques, and are thought to be highly repetitive and gene-poor). The gene-rich areas make up euchromatin. Gene numbers are taken from the original sequence publications. Most numbers have since changed slightly and different sources give different estimates depending on protocols.

The genome of humans and of other organisms was sequenced
The age of multilaboratory collaboration was established

Two features seem to dominate this period. The first is the sequencing of the DNA of several organisms including humans. One simply now had the possibility to develop techniques that permitted determining the nucleotides that were present along a whole genome, as complex as that of humans, with its 46 chromosomes and 3 billion bases. The second event was facilitated by the creation of the European Union which led to the formation in Brussels of international bodies that supported collaboration between laboratories in countries that belonged to the Union and outside its frontiers. The consequence was a collaboration covering dozens of laboratories and scientific reports including over 100 scientists. This was a new situation dictated by the technological needs of rapidly sequencing very long DNA sequences which demanded the simultaneous efforts of many colleagues.

1. Sequencing comes of age
The first nucleotide sequence of the DNA of a chromosome of a eukaryote was published in 1992. The chromosome III of yeast, one of the smallest, was found to be 315,357 base pairs long as a result of the joint effort of 147 scientists working in 35 different laboratories. This chromosome contained 182 open reading frames, *i.e.* structural gene sequences flanked by an initiation and a termination codon. Surprisingly as many as 80% of these genes showed no obvious homology to other yeast genes studied previously (S.G. Oliver et al. 1992). The Huntington's disease gene was sequenced by a group comprising 57 researchers. Of significance was that a short sequence (a trinucleotide) increased in number in the individuals having this disease (M.E. MacDonald et al. 1993). Later another large group of 108 scientists from 29 European laboratories, sequenced the yeast chromosome XI. This was larger, with 666,448 base pairs. Some of the 331 open reading frames had introns (B. Dujon et al. 1994).

By 1995 40 colleagues described the sequence of the bacteria *Haemophilus influenzae* and in the same year 29 research workers completed the sequence of *Mycoplasma genitalium* (R.D. Fleischmann et al. 1995, C.M. Fraser et al. 1995). A technique was developed that shortened appreciably the time necessary to sequence human chromosomes. The result was the publication of the nucleotide sequence of chromosome 22 (I. Dunham et al. 1999). The results from the sequencing of eukaryotic genomes were summarized by P. Bork and R. Copley (2001).

2. Human disease
Human genes were now the target of intensive research especially those connected with disease.

The fragile site of the human X chromosome turned out to contain a CGG short sequence that increased in number and was connected with mental retardation (A.J.M.H. Verkerk et al. 1991). The organization of the gene that was implicated in Marfan's syndrome was studied by L. Pereira et al. (1993).

A gene responsible for kidney disease in humans was analysed. This time it was the product of the gene, the protein, that was sequenced, which was found to contain 4,320 amino acids (J. Hughes et al. 1995). The Alzheimer's disease gene was located in the human chromosomes. Mutations in this gene were responsible for the onset of the disease (R. Sherrington et al. and G.D. Schellenberg 1995).

3. Cancer and aging
A problem connected with cancer was aging. Inroads into this area were made using molecular tools. W.C. Orr and R.S. Sohal produced transgenic lines in *Drosophila* with extra copies of the genes for superoxide dismutase and catalase. These flies aged at a slower rate (1994). Another phenomenon that is related to both aging and cancer is the behaviour of telomeres. N.W. Kim et al. found that human somatic tissues lacked telomerase activity. On the contrary normal germ cells and cancer cells had a high degree of telomerase activity (1994).

The gene connected with obesity in the mouse was cloned by Y. Zhang et al. (1994) and human anti-oncogenes, such as BRAC1, were found to be involved in breast and ovarian cancer (Y. Miki et al. 1994).

4. Development
M.C. Mullins and C. Nüsslein-Volhard (1993) who had studied development in *Drosophila*, now searched for a vertebrate where they could dissect development genetically, by producing hundreds of mutations in the zebra fish. A result in the same direction was obtained by R.J. Bollag et al. (1994) who showed that the *T* genes of the mouse affected development both in invertebrates and vertebrates.

G. Halder et al. (1995) found in *Drosophila* a master gene that controlled the formation of the eye. The next year two research groups found that a homeobox protein was able to bind to specific messenger RNAs (J. Dubnau and G. Struhl 1996, R. Rivera-Pomar et al. 1996).

5. Chromosome organization

One of the typical examples of the view that most events occurring in the chromosome were random phenomena, had been, for many years, crossing over. However, this phenomenon turned out to be ordered when R.M. Story, I.T. Weber and T.A. Steitz (1993) established the three-dimensional structure of RecA in *E. coli*. This was a protein which had a major role in the guidance of crossing over as well as of DNA repair. Both phenomena appeared intimately connected at the molecular level and both appeared to be highly ordered. DNA repair was the guardian of the original order of the chromosome which ensured that errors were eliminated and crossing over was the guardian of diversification that ensured that genes were recombined in innovative forms.

Another example of the determination of the behaviour of the chromosome by specific proteins was found in the *mei-S332* gene of *Drosophila* which appeared to be responsible for the maintenance of the union of sister chromatids during the first division of meiosis (A.W. Kerrebrock et al. 1995).

The size of the structural gene, which for a long time had been a source of conjecture, could now be calculated with more accuracy. As many as 90 *Drosophila* genes, that were active in transcription, had sizes ranging from 319 to 4,749 base pairs (G. Maroni 1993).

6. The cell organization

Centrosomes were found to contain gamma-tubulin complexes which functioned as origins of microtubules. Two new organelles were discovered in oogonia and cytoblasts of *Drosophila*: 1) *Fusomes* were gelatinous masses that were found in the ring canals that made connections between cells during cystocyte divisions. 2) *Spectrosomes* were spherical structures present in the cytoplasm of these cells that associated with the spindle and participated in cell division (M. Moritz et al. 1995).

7. Chromosome evolution

Chimpanzees, gorillas and orangutans have 24 pairs of chromosomes whereas humans have only 23. J.W. Ijdo et al. (1991) sequenced the region of chromosome 2 in humans which was supposed to be responsible for the fusion of two ancestral chromosomes into the human V-shaped chromosome 2.

All known protein kinases turned out to have a common catalytic core. This was one more result putting in evidence the conservation of structures and genes throughout evolution (D.R. Knighton et al. 1991).

A mutation in cystic fibrosis genes could be traced to have occurred in European populations 50,000 years ago (N. Morral et al. 1994).

C. Bult et al. (1996) studied the genome of an archaeon finding its genes not to be related to those of other species.

Nucleotide sequencing of mitochondrial DNA of humans from different continents and of great apes led to the conclusion that all human mtDNA molecules had their origin in an African woman who lived about 140,000 years ago (S. Horai et al. 1995) but these results ought to be considered preliminary.

8. Neurobiology

The search for genes involved in memory led to their isolation in *Drosophila* (T. Tully et al. 1994). A gene that encoded a leptin receptor was transcribed in the hypothalamus region of the vertebrate brain (L.A. Tartaglia et al. 1995).

9. Biotechnology

The first case of gene therapy was obtained by W.F. Anderson (1990) who performed an experiment in a patient with an adenosine deaminase deficiency. White blood cells were incubated with a retroviral vector which carried a normal gene which produced the missing enzyme. The cells transformed in this way were injected into the patient and corrected the deficiency.

M.E. Gurney et al. (1994) transferred human genes into mice. These genes were the mutant forms of superoxide dismutases. The transgenic mice synthesized the mutant enzymes and got paralyzed in a way similar to humans carrying the disease.

The event that filled the headlines of the international press was the cloning of the sheep "Dolly" (I. Wilmut et al. 1997). This led to the proposal, by various laboratories, that humans could be cloned. It was also suggested that the production of humans, without heads, could be used as organ donors (1998).

This period closed with two events significant for the future of chromosome research and molecular biology. A flowering plant, *Arabidopsis*, which started to be studied genetically as early as the 1940s, became the object of a full DNA sequencing of its genome completed in 2000 by The Arabidopsis Genome Initiative. But the highlight was the publication of the DNA sequencing of the human genome by two separate groups: The International Human Genome Sequencing Consortium and J.C. Venter et al. (2001).

Ninth Period 2001–2010

REPRESENTATIVE GOALS OF THE 2010 PROJECT

1- to 3-Year Goals

Develop essential genetic tools, including the following:
- comprehensive sets of sequence-indexed mutants, accessible via database search
- whole-genome mapping and gene expression DNA chips
- facile conditional gene expression systems.

Produce antibodies against, or epitope tags on, all deduced proteins.

Describe global protein profiles at organ, cellular, and subcellular levels under various environmental conditions.

3- to 6-Year Goals

Create a complete library of full-length cDNAs.

Construct defined deletions of linked, duplicated genes.

Develop methods for directed mutations and site-specific recombination.

Describe global mRNA expression profiles at organ, cellular, and subcellular levels under various environmental conditions.

Develop global understanding of posttranslational modification.

Undertake global metabolic profiling at organ, cellular, and subcellular levels under various environmental conditions.

10-Year Goals

Plant artificial chromosomes.

Identify *cis* regulatory sequences of all genes.

Identify regulatory circuits controlled by each transcription factor.

Determine biochemical function for every protein.

Describe three-dimensional structures of members of every plant-specific protein family.

Undertake systems analysis of the uptake, transport, and storage of ions and metabolites.

Describe globally protein-protein, protein-nucleic acid, and protein-other interactions at organ, cellular, and subcellular levels under various environmental conditions.

Survey genomic sequencing, and deep EST sampling from phylogenetic node species.

Define a predictive basis for conservation versus diversification of gene function.

Compare genomic sequences within species.

Develop bioinformatics, visualization, and modeling tools that will facilitate access to all biological information about a representative virtual plant.

SOMERVILLE and DANGL 2000 -
Goals of the *Arabidopsis* project.

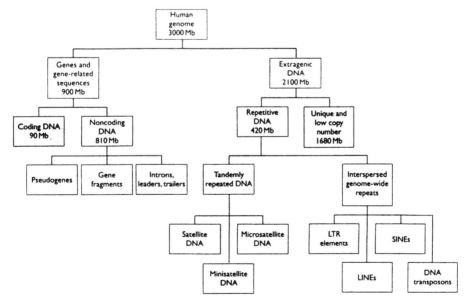

BROWN 1999 -
The organization of the human genome
based on STRACHAN and READ (1996).

The post-genome era
The task that lies ahead

The mapping of the human genome is a landmark in the development of genetics and molecular biology. It has resulted in the identification of most base pairs that build the DNA of the 46 chromosomes of humans. These bases are about 3 billion. Suddenly we are confronted with a highway which extends along a colossal stretch of molecular information when seen at the Angstrom level of the component atoms of the DNA bases.

Genetics was born 100 years ago, when the rules of inheritance were rediscovered in plants in 1900. Thus, it took a century to assemble observations and experiments that led to the building of a scientific edifice that finally culminated in this technical achievement.

It ought to be realized that the mapping of the genome is in itself a product of technology. It was mainly achieved by using machines, to increase the speed of the mapping process. Constant improvement in sequence automation now allows reading 500 base pairs in 50 minutes. Hence it represents a situation different from the proposal of a model, such as that of Watson and Crick, or the formulation of a new interpretation that would predict the behaviour of the chromosome as a unit. Moreover, at first sight, it seems that genetics has achieved its main goal, and that we are now faced with the easy task of finding where the genes are and what they do. However, on the contrary, it is now that the difficult task begins, since we will be in a position to get more exact answers to many questions that earlier could not even be formulated.

Let us consider some of the many molecular and atomic situations that we will be led to investigate in the near future.

The shifting properties of the DNA and RNA molecules
It has been known since the early studies of DNA chemistry (1950s) that the configuration of this molecule changes with the water content (A, B and other forms). The amount of water in cells varies appreciably from tissue to tissue. The grooves of the DNA molecule are different in these forms and the exposed base configuration is expected to affect the binding of proteins involved in gene expression. But, besides the bases, the phosphate groups of the DNA backbone are now known to play an active role in protein-DNA recognition. There is an interaction between the positive charges of the amino acids of the protein and the negative charges of the phosphate component (G.D. Stormo and D.S. Fields 1998).

There are also in DNA interactions between base pairs that add stability to the double helix. These have been called stacking forces and have an effect at a distance along the DNA molecule.

RNA also has peculiar properties. As predicted by F.H.C. Crick, the recognition between messenger RNA and transfer RNA does not necessarily follow the usual rules of base pairing, but codon recognition may vary. The triplet of nucleotides is slightly curved and as a result new types of base-pairs occur. Instead of the regular adenine-uracil bonding, uracil-guanine pairs are permitted. Inosine can also base-pair with adenine, cytosine and uracil. This versatility can even increase in some cases extending to all four bases (a phenomenon called superwobble) (T.A. Brown 1999).

Thus, DNA and RNA have shifting properties that may lead to novel interactions that are now starting to be discerned. The atomic properties of the molecules that build up DNA and RNA seem to be a promising area of research.

Histones are obligatory components of eukaryotic chromosomes and are regulators of their function
Any stretch of DNA by itself does not make a eukaryotic chromosome. In organisms with a nucleus the DNA is regularly associated with proteins. This is an event that started with the protozoans and that has extended to higher organisms including humans (a span of 2.5 billion years).

The main proteins associated with DNA are the histones, well studied and isolated since the 1960s. The coupling histone-DNA builds the basic unit of chromosome coiling – the nucleosome. This unit has been visualized in the electron microscope and its components have been disassembled and self-assembled again.

In the last 30 years it has been known that this molecular complex is of paramount importance in the expression of specific genes. Hyperacetylation of histones is generally correlated with transcriptionally active chromatin whereas hypoacetylation is associated with transcriptional silencing. This phenomenon depends on the limited access to the DNA of activating factors and the binding of repressors to promoter DNA (U. Mahlknecht et al. 2000).

Thus, the function of a gene is not only dependent on its base sequences but on the way these sequences are bound to histones. We simply do not know what physico-chemical mechanisms make the histones attach to certain genes and not to others, during the various stages of chromosome reorganization occurring during the cell cycle. Moreover additional types of histones are being discovered such as the linker histones which are located between nucleosomes and the core histones present at other sites not yet precisely identified (E. Pennisi 1996).

We also tend to forget the structural proteins
Hardly ever mentioned are the proteins called structural. These are a different class that derives its name from the fact that using appropriate chemical treatments one can remove the DNA and the histones from a chromosome and its thread-structure remains apparently intact (A.E. Mirsky and H. Ris 1949, J.R. Paulson and U.K. Laemmli 1977). This seems astonishing because of our mental fixation on the mighty importance of DNA. The structural proteins seem to be a kind of scaffolding of the chromosome body. The type of molecular interactions that occur between the structural proteins, the histones and the DNA remains highly obscure.

New types of proteins are being discovered
As the DNAs of the centromere and the telomere became sequenced their proteins were also found to be different from those of the chromosome arms. Seven proteins, are present in the centromere and one of them CENP-A, is similar to histone H3. Special telomeric proteins build a proteinaceous cap at the chromosome end but they remain to be identified in detail (O. Vafa and K.F. Sullivan 1997).

The RNA of the chromosome – a multifunctional molecule
The total RNA mass of a cell consists of approximately 80 to 85% ribosomal RNA. Transfer RNA, small nuclear RNA and low molecular weight species make up 10–20% while messenger RNA represents only 1–3%. All these RNAs arise in the chromosome as nascent structures transcribed from specific DNA sequences. But they soon leave the chromosome moving to the nuclear sap or to the cytoplasm. Only in a few instances and under special conditions are they part of the chromosome. The ribosomal RNAs become retained in the nucleolus, which is a partly independent organelle of the chromosome. Other RNAs are an integral part of the chromosome without which DNA cannot function. In eukaryotes DNA cannot start its own replication alone. It needs the intervention of an RNA primer that is later discarded. Telomere replication cannot take place either without the presence of a specific RNA. This replication is carried out by the telomerase enzyme but this consists both of protein and RNA which is used as a template (J. Lingner et al. 1997, E.H. Blackburn 2000).

RNA is a "jack-of-all-trades". All the time new functions are found for this molecule. Practically every genetic event in the cell could occur using only RNA, since it can be: the genetic material (as in certain viruses), the messenger, the transfer, the ribosomal, but above all it can even function as a protein (as in a ribozyme). It could be stated that this is the molecule of the future since it may have stood alone for the ancestral emergence of the cell.

The chromosome is a well delimited unit
So far we have dealt with the molecules that build the chromosome as it works in the cells of higher organisms. Now let us consider the way the bases of the DNA have become organized into functional units building not only minor and major aggregates but different types of signal segments that delimit them structurally and functionally. Only by understanding the total organization of the chromosome will we be able to predict its behaviour as a unit.

Any chromosome seen under the microscope at metaphase of mitosis is a well defined structure. It usually appears as a cylindrical rod sharply delimited from the adjacent cytoplasm. Few cell organelles are so precisely terminated as the chromosome. This is achieved by specialized regions recognized at the microscopic level and which have now been defined in molecular terms. It may be said that the centromere contains the start signal of the chromosome and the telomere its stop signal since, in many species, chromosomes consist of a single arm, that originates in a centromere and terminates in a telomere. Most chromosomes have two arms but there are no chromosomes with a star configuration. Following X-rays and other treatments some star forms have been found, which means that they can occur (P. Slijepcevic et al. 1997). What mechanism has hindered the radial pattern remains unknown.

The initiation sequence of a chromosome – the centromere
The centromere is the region that leads the chromosomes to the poles. This function is obviously primary, because without it, chromosomes cannot move on the spindle and as a consequence cannot be transferred to daughter cells. The centromere is not only a primary site, but equally relevant is that the most important properties of a chromosome have their origin in this region. It has been known for over 40 years that crossing over, chiasmata, structural rearrangements, gene territory, chromomere patterns and other genetic phenomena show a distribution originating at the centromere or directly related to its position. Thus it does not come as a surprise that the DNA of the centromere is of a special type. It is made of repetitive units which in human chromosomes are 171 base pairs in length and are called *alphoid* DNA. About 500,000 bases of this DNA are needed for complete centromere function. Unexpected is the finding that associated with this DNA there are seven proteins not found elsewhere in the chromosome (J. Carbon 1984, A.F. Pluta et al. 1995).

The centromere appears as a highly sophisticated

region not only functionally but also molecularly, and one wonders what secrets it carries that, once unveiled, will allow us to understand its major function as the determinant of the chromosome's main properties. The centromere apparently builds a field of action which controls the behaviour of the chromosome as a unit.

The termination sequence of a chromosome – the telomere

Radiation effects, that break chromosomes into pieces, showed from the start, that such chromosome segments even if they had a centromere could not survive. Only when they had their natural ends attached to the extremities, did they become genetically stable. The natural end was called a telomere (end region). Its main function is a stop signal that locks the extremity of the chromosome. Chromosomes may vary extremely in length. In birds the mini-chromosomes are near the resolution of the light microscope (0.2 microns) but other chromosomes such as those of rye, at pachytene, may measure 100 microns. Moreover one single chromosome of the plant *Tradescantia* contains more DNA than the 23 chromosomes of humans. The question then arises: What mechanism locates the stop signal after only a few DNA sequences or after a huge DNA array that may contain over one billion bases of DNA? At present we have no idea.

Like centromeres, telomeres also influence the location of chromosome properties in their vicinity (A. Lima-de-Faria 1980). Since then they have been found to repress the transcription of genes located nearby (D. De Bruin et al. 2000). Telomeres have been sequenced and their DNA is similar from protozoans to humans. In this species telomeric DNA is made up of hundreds of copies of a repeated motif 5′-TTAGGG-3′ with a short extension. Again, like at the centromere, there are special proteins that build a proteinaceous cap that protects the ends. Here too, there are many concealed features to be discovered before we understand the formation of the stop process and the long range effect that the telomere seems to have on genes located throughout the arm.

Superfluous RNAs became essential RNAs

The large collection of RNAs that is present in a cell has increased all the time. At present four classes of RNAs are distinguished: 1) Those involved in processing and modification of transfer, ribosomal and messenger RNA. 2) Those which regulate gene expression. 3) The RNAs that participate in "housekeeping functions" such as the telomerase RNA and 4) RNAs with no established role in cellular metabolism, such as 7SKRNA.

From the beginning these and other RNAs were considered superfluous but now it is known that they are instead essential. A recent unanticipated discovery was the "minor spliceosome". It contains U11 and U12 RNAs, that when they were originally identified, were believed to have no function. They are now known to participate in the removal of "minor" introns. Two other new RNAs are U4 and "U6 atac" which were discovered in connection with spliceosome research (W.-Y. Tarn and J.A. Steitz 1997, M.G. Caprara and T.W. Nilsen 2000).

Many well characterized RNAs still have no established role in cell metabolism and are waiting to have their function identified.

Superfluous DNAs became essential DNAs

It became evident, already during the 1970s, that most DNAs had functions which were not directly related to the production of proteins. The accumulated data disclosed the following: (1) There were satellite DNAs consisting of as much as 1 million copies that were formed by short nucleotide sequences which could not give rise to a polypeptide chain of the size present in most proteins. (2) Nuclear RNAs were exceedingly large molecules that later were chopped down to the small messengers of proteins. In other words, a large part of the original DNA sequence was not used to build the protein. (3) Heterochromatin, that was known to have little or no genetic activity, was found to contain the highest amount of DNA present in chromosomes. These, and other data, led to the conclusion that a large fraction of the DNA present in a genome was not implicated in the formation of structural genes.

Later, split genes were discovered, which consisted of introns and exons. The exons code for the protein but the introns do not have this function. The view was then expressed that introns ought to be small and insignificant segments whereas the exons should be the larger segments since they had this important function. It turned out to be exactly the reverse. Introns are on the average 10 times larger than exons. Moreover introns, can also change their function, participating in protein formation. Subsequently DNA satellites were found to be of importance, as regulatory agents of gene function, some of them being involved in cancer. Other repetitive sequences are crucial in maintaining the framework of the chromosome. To this list can be added the pseudogenes, the transposons, the microsatellites and the minisatellites (T.A. Brown 1999).

This situation may become better elucidated when the average size of the structural gene as well as the actual number of genes are established.

The number of human genes has successively decreased

From the number of proteins present in the human body it was generally agreed in the 1980s that we should have between 200,000 and 100,000 structural genes.

In 1996 more than 4,000 human genes had been identified, and many had been sequenced, which allowed estimating the average size of a human gene. This varied, in thousands of nucleotides, from 1.5 to 300 (in one case it exceeded 2,000). Their number of introns varied from 2 to more than 50. From the estimation of the average size of a human gene and a gene number of 70,000 to 100,000, several authors came to the conclusion that only 3% of the total human genome consisted of structural genes (T. Strachan and A.P. Read 1996).

The sequencing of human DNA led to the conclusion that humans had only 39,114 genes (P. Bork and R. Copley 2001). This is a dramatic reduction that puts us close to the plants and the worms (25,498 and 19,099 genes respectively). These findings raise many questions concerning gene permanence as well as gene diversification. The organization of the human genome was summarized by T.A. Brown (1999) based on information provided by T. Strachan and A.P. Read (1996).

Goals that may be considered in the coming years
C. Sommerville and J. Dangl (2000) have summarized in a Table the 1- to 3-year, as well as 3- to 6-year and 10-year goals representative of the *Arabidopsis* project. This may in part apply to the work being carried out on other species.

Their interest is focused on: 1) Gene expression. 2) Antibodies against deduced proteins. 3) Protein profiles at the organ and other levels. 4) Methods for directed mutations. 5) Posttranslational modification. 6) Regulatory circuits controlled by transcription factors. 7) Protein-protein and protein-nucleic acid interactions. Their Table contains more items, but these suffice to give a picture of some of the directions that research may follow in the coming 10 years.

PART II

THE TECHNOLOGY THAT ALLOWED THE STUDY OF THE CHROMOSOME: 1900 TO 2001

From staining methods to DNA sequencing

It was during the 1910s and 1920s that the study of chromosomes became of general interest. To facilitate microscopic observation the cells had to be available as thin layers since the object had to be sufficiently transparent to the incident light rays. The task was made easier by adapting the technology used in microbiology to chromosome research. For this purpose a drop of blood was poured on a slide. Another slide was used to spread it, obtaining in this way a single cell layer. Another way of obtaining a tissue smear was by spreading the cells with the help of a cover slip (M. Langeron 1934). But the most used technique became the embedding in paraffin of an organ, that was successively mounted on a block, and sectioned with a microtome (A. Guilliermond and G. Mangenot 1941, G.L. Humason 1972). The chromosomes remained in their natural position but were often mutilated or lost. To preserve the original structure, the tissues were immersed in several fixatives, or their combinations, before being embedded in paraffin (J.R. Baker 1950). Staining procedures were selected that coloured mostly the chromosomes, leaving as much as possible the cytoplasm as a transparent mass, otherwise the chromosomes could not be easily distinguished from an irregular background.

The large field binocular microscope was employed in the initial stages of dissection of the material (M. Langeron 1934). Soon the high resolution binocular microscope was introduced, which minimized the fatigue resulting from many hours of cell analysis (L.C. Martin and B.K. Johnson 1949). The inclusion of oil immersion objectives allowed a much better resolution of the chromosome and the photographic camera was mounted on the top of the microscope. However, the photographs obtained were far from satisfactory. The chromosomes were not on a single plane, part of them disappeared from the focus. For this reason the drawing of complete chromosomes was an obligatory complement, to obtain a reliable representation of the structure of a single chromosome or of the chromosome complement of a cell. The introduction of the camera lucida, which used a prism and a mirror, which projected the chromosome on a piece of paper, became a most reliable method (M. Langeron 1934). In the following years the staining of chromosomes improved, resulting in fine preparations (C.D. Darlington and L.F. LaCour 1960).

One of the problems was the avoidance of artifacts. It was necessary to be sure that the chromosomes observed under the microscope, as coloured structures, were a correct representation of their condition in the living state. This problem was solved when chromosomes, and the whole cell division, were analyzed in living cells (E. Strasburger 1924) and when it became possible to follow under the microscope the effect of fixatives demonstrating that they did not alter the original form and structure of the chromosomes (K. Belar 1928).

The discovery of the giant chromosomes, in the salivary glands of *Drosophila*, led to the development of a simple technique in which the larvae were dissected under the large field binocular microscope, the glands being removed, fixed and stained to analyze the pattern of their bands (M. Demerec and B.P. Kaufmann 1961).

Soon the ultra-violet microscope was invented with its lenses made of quartz. This was an expensive and complex instrument, but it allowed distinguishing the location of RNA and DNA in the cell from that of protein, due to their wavelength absorption. DNA was shown to be mainly present in the chromosomes, whereas RNA was to be found in the nucleolus and the cytoplasm. These results heralded the advent of cell chemistry (T.O. Caspersson 1936). Colchicine turned out to arrest cell division without affecting the doubling of the chromosome number of a cell. The use of this drug permitted the experimental production of polyploids, a technique which had consequences for plant breeding by producing new varieties (O.J. Eigsti and P. Dustin 1955).

The discovery of the phase microscope was not so dramatic, and its impact had no long range effects. However, it became a most useful tool in tissue culture (J. Brachet 1957). Another technique led to the mass isolation of chromosomes (A.R. Gopal-Ayengar and E.V. Cowdry 1947), which allowed obtaining large amounts of material. In this way the proteins, that were associated with DNA in chromosomes – the histones – were well identified (D.M. Fambrough 1969).

Note: As mentioned in the *Introduction* the figures selected to exemplify a given technique, experiment or observation, were not always chosen from the works in which they were first described. The pioneering works have usually been mentioned in the chapters in which the subject has been treated in detail. Instead publications were used, in which the findings were clearly depicted or represented in a simplified form. This criterion was followed in the choice of the figures included in most plates.

After the Second World War (1939–1945) two new techniques were applied to the study of chromosomes that had an enormous impact. These were: electron microscopy and ultra-centrifugation.

The early models of the electron microscope were clumsy machines (L.C. Martin and B.K. Johnson 1949). At that time the specimen preparation was a procedure consisting of many steps difficult to control. Another permanent source of error was the preparation of glass knives which were obtained by breaking thick plates of glass with strong pliers. The sections had to be ultra thin otherwise the electron beam would not penetrate them (D.C. Pease 1960). This technique revealed itself most fruitful for the analysis of the structure of the cytoplasmic cell organelles, but became a disappointment in what concerned the metaphase chromosomes (A.T. Sumner 1977). These turned out to be too thick and compact for the electron beam. Initially the electron microscope did not add new data to chromosome organization, that was not known earlier from the light microscope. On the other hand, the nuclear envelope, the chloroplasts, the mitochondria and the endoplasmic reticulum suddenly appeared as beautiful structures full of novel details displaying a remarkable organization at the E.M. level.

Very simple hand centrifuges had already disclosed that the cell components could be stratified into distinct layers (M. Langeron 1934, N. Andresen 1942). But it was the introduction of the ultra-centrifuges, with their enormous speeds, that resulted in a clear-cut stratification of cell organelles, into well defined bands, that could be collected from plastic tubes as highly pure fractions (S. Osawa and A. Sibatani 1967, J.R.B. Hastings 1972, H. Lodish et al. 1995).

The chromosome continued to be too voluminous for electron microscopy but its nucleic acid components: the RNA and DNA molecules, which could now be separated biochemically, were small enough to give splendid pictures in the E.M., even revealing the areas in which the strands of these molecules were paired or remained separated (M. Thomas 1967, D.S. Hogness et al. 1975, D.M. Glover 1984).

Radioisotopes had been available for a long time, but the length of the tracks made by their electrons, were too long to allow a resolution compatible with chromosome size. The first essays with P^{32} showed that one could distinguish between two groups of chromosomes, separating after cell division, but not between single chromosomes or specific chromosome segments (A. Howard and S.R. Pelc 1951). Moreover the use of liquid emulsion to detect the presence of the emitted electrons resulted in an uneven picture of particle distribution. The breakthrough came with the introduction of molecules marked with tritium and the use of stripping film (G.A. Boyd 1955). Electrons from tritium travelled on the average only 1 to 2 microns and stripping film consisted of an even emulsion. The analysis of the replication of DNA in single chromosomes suddenly became possible (J.H. Taylor 1958, A. Lima-de-Faria 1959).

At the same time biochemists looked for ways of identifying the amino acids of proteins (G.L. Brown et al. 1950, H.D. Springall 1956). This was later followed by the sequencing of these molecules. But this development became of more importance for understanding cell function at the molecular level, than to clarify chromosome organization. In the meantime crystallographers were able to detect the structure of DNA by X-ray diffraction (R.E. Franklin and R. Gosling 1953). The discovery of its helical structure became the basis of the DNA model.

Since the turn of the century chromosomes were known to have a spiral structure, but they continued to conceal their inner differentiation and mitotic chromosomes were difficult to distinguish from one another. Suddenly it turned out that different chemical treatments led to band formation along the chromosome body. Moreover the pattern of these bands turned out to be specific for every chromosome. In species with many chromosomes such as the human complement, it was most difficult to distinguish with accuracy similar chromosomes. The banding procedure solved this problem (A.F. Dyer 1979, A.A. Sandberg 1980, C.W. Yu 1980).

After the sequence of amino acids had been established for several key proteins the search was made for a technique that would allow sequencing the bases in RNA and DNA. Soon several methods became available. This led to the sequencing of the bases in nuclear genes as well as in the DNAs of mitochondria and chloroplasts (K. Arms and P.S. Camp 1995).

The strong preoccupation with the organization of the gene and of DNA, had for many years relegated the problem of embryonic development to a secondary position in genetic research. By the late 1950s the role of the nucleus in cell differentiation became an acute problem. Experiments led to the introduction of nuclei from somatic cells into eggs that lacked their own nucleus. Also the fusion of embryos from parents with different genetic markers was obtained with a view of following their path in development. These studies had implications for the understanding of cancer (B. Mintz 1971). Later the use of *Drosophila* in the study of development, allowed establishing a direct relationship between specific genes and body compartments (A. Garcia-Bellido et al. 1979).

Two techniques that facilitated the emergence of bioengineering were the discovery of restriction enzymes and the separation of nucleic acid segments by agarose gel electrophoresis (E.M. Southern 1975, K. Arms and P.S. Camp 1995). Restriction enzymes turned out to be highly specific, cutting DNA at well defined base sequences that built palindromes (D.S. Hogness et al. 1975). The short segments of DNA obtained after the digestion with such enzymes could be separated by electrophoresis, according to their size, in a clear and reproducible way (R.W. Old and S.B. Primrose 1980). This technology became the key process for the combination of nuclear gene sequences with DNAs from virus particles. This allowed the transfer of these new gene combinations into bacterial and mammalian cells. Another technique led to the hybridization in situ with RNAs which were labelled with tritium, but lately one has used biotinylated probes (H. Scherthan et al. 1998). A closer approach to gene function was obtained when O.L. Miller and A.H. Bakken (1972) developed an electron microscopy technique which allowed visualizing the transcription process in ribosomal genes.

Later developments have seen the commercial development of instruments that sort out cells marked in different ways with fluorescent dies as well as the separation of specific chromosomes (D.R. Parks and L.A. Herzenberg 1982, T.A. Brown 1999). This procedure is carried out at high speeds, allowing an effective localization of genes in chromosomes.

Procedures are now available that permit the extraction of amniotic fluid during pregnancy, a valuable technique in chromosome diagnostics (T. Friedman 1971). Other methods permit to inject transformed cells into patients suffering from specific genetic defects or from cancer (D.M. Glover 1984, K. Arms and P.S. Camp 1995). Thanks to their engineered DNA these cells can be addressed to specific organs where the defective genes are harboured. This technique has been in effect since 1990 and over 100 patients have been treated.

Two techniques, used mainly in the last decade, that led to startling results in the field of chromosome research, were the polymerase chain reaction (PCR) and the sequencing of DNA. By using an enzyme that tolerated high temperatures it was possible to obtain large numbers of copies of selected DNA segments (R.K. Saiki et al. 1985, B. Alberts et al. 1994). The sequencing of DNA was improved by the construction of machines that simplified this process. This resulted in a rapid mapping of the DNAs of a diversified number of species which included humans (J.C. Venter et al. 2001).

THE TECHNOLOGY THAT ALLOWED THE STUDY OF THE CHROMOSOME -- 1
1900 TO 2001.

MARTIN and JOHNSON 1949 - Early model of the binocular microscope.

LANGERON 1934 - Microscope with attached camera lucida.

MARTIN and JOHNSON 1949 - Microscope with attached photographic equipment.

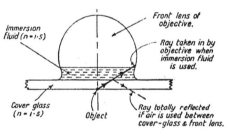

MARTIN and JOHNSON 1949 - The immersion oil objective which allowed sharper pictures of chromosomes.

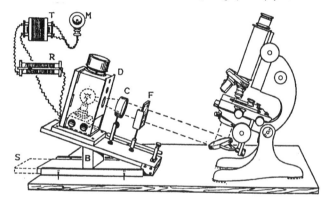

DARLINGTON and LA COUR 1960 - Simple equipment was used in the study of chromosomes, as late as 1960.

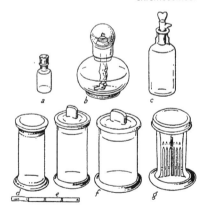

DARLINGTON and LA COUR 1960 - Glass implements used in the staining of chromosomes.

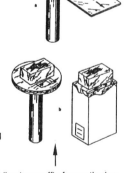

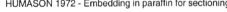

HUMASON 1972 - Embedding in paraffin for sectioning.

GUILLIERMOND and MANGENOT 1941 - Microtome for sectioning tissues embedded in paraffin.

THE TECHNOLOGY THAT ALLOWED THE STUDY OF THE CHROMOSOME -- 2
1900 TO 2001.

LANGERON 1934 - Spreading of cells.

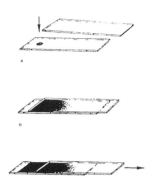
HUMASON 1972 - Preparation of single cell layers.

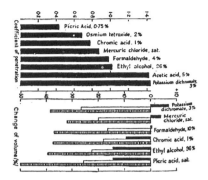
BAKER 1950 - Properties of most used fixatives.

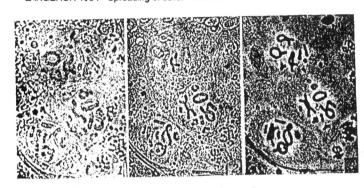
BELAR 1928 - Successive stages of the action of fixatives on living material.

GOPAL-AYENGAR and COWDRY 1947 - Mass isolation of chromosomes.

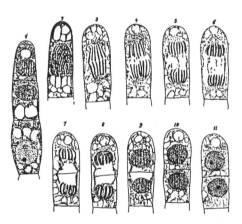
STRASBURGER 1924 - Chromosomes seen in living cells.

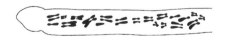
EIGSTI and DUSTIN 1955 - Colchicine treated chromosomes.

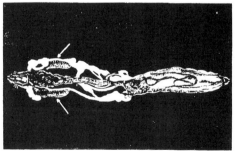
DEMEREC and KAUFMANN 1961 - Dissection of *Drosophila* larvae under the large field binocular microscope.

THE TECHNOLOGY THAT ALLOWED THE STUDY OF THE CHROMOSOME -- 3
1900 TO 2001.

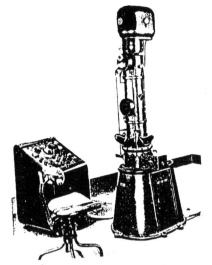

MARTIN and JOHNSON 1949 - An early model of the electron microscope.

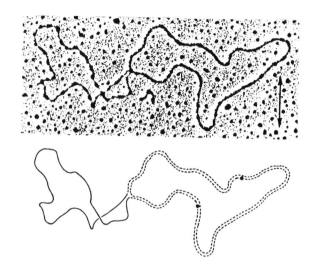

HOGNESS et al. 1975 - DNA molecules in the E.M. partly single-stranded and partly double-stranded.

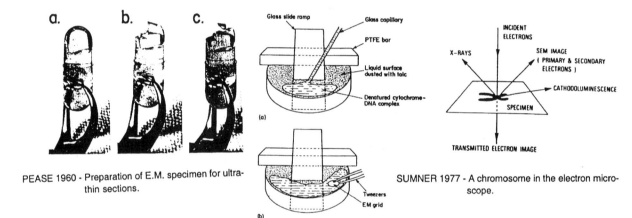

PEASE 1960 - Preparation of E.M. specimen for ultra-thin sections.

GLOVER 1984 - Preparation of DNA molecules for the E.M..

SUMNER 1977 - A chromosome in the electron microscope.

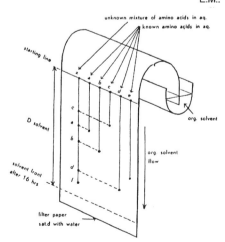

SPRINGALL 1956 - Separation of amino acids.

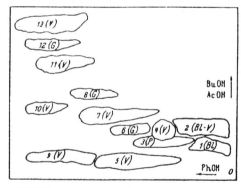

BROWN et al. 1950 - Identification of amino acids in proteins.

THE TECHNOLOGY THAT ALLOWED THE STUDY OF THE CHROMOSOME -- 4
1900 TO 2001.

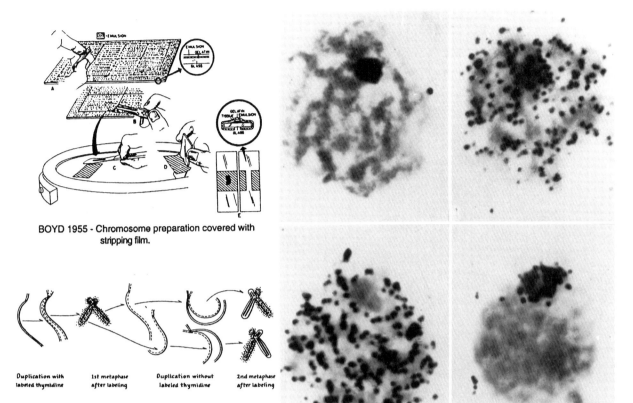

BOYD 1955 - Chromosome preparation covered with stripping film.

TAYLOR 1958, 1963 -
Semiconservative replication of eukaryotic chromosomes in *Vicia faba*. Diagrammatic representation of the distribution of DNA strands during chromosome duplication. Broken lines represent labeled strands; unbroken lines represent unlabeled strands. The dots represent grains in the autoradiographs.

LIMA-DE-FARIA 1959 -
Late replication of heterochromatin. Autoradiography of the chromosomes of *Melanoplus* at pachytene, labelled with tritiated thymidine. Upper left, unlabelled cell showing heterochromatic X chromosome (dark body) and euchromatic autosomes. Upper right, all chromosomes labelled. Lower left, autosomes labelled and X unlabelled. Lower right, only X labelled.

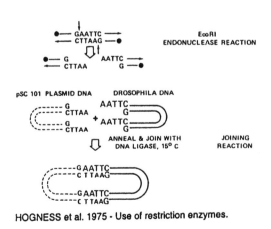

HOGNESS et al. 1975 - Use of restriction enzymes.

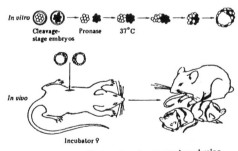

MINTZ 1971 - Fusion of embryos produced mice chimaeras.

THE TECHNOLOGY THAT ALLOWED THE STUDY OF THE CHROMOSOME -- 5
1900 TO 2001.

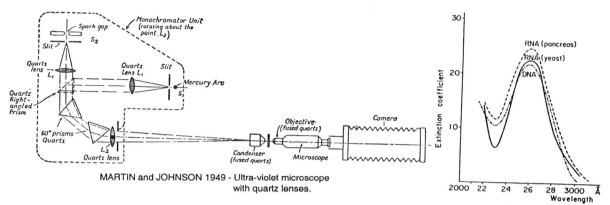

MARTIN and JOHNSON 1949 - Ultra-violet microscope with quartz lenses.

CASPERSSON 1936 - Wavelength absorption of RNA and DNA.

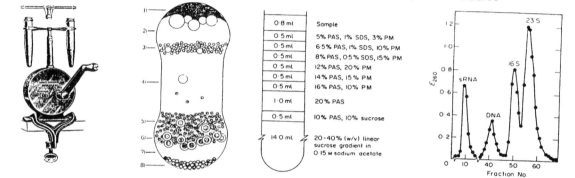

LANGERON 1934 - A simple table centrifuge.

ANDRESEN 1942 - Stratification of the components of a cell into layers by centrifugation.

HASTINGS 1972 - Sucrose gradient centrifugation allowed a fine resolution of cell components.

OSAWA and SIBATANI 1967 - Fractionation of RNA and DNA into distinct peaks.

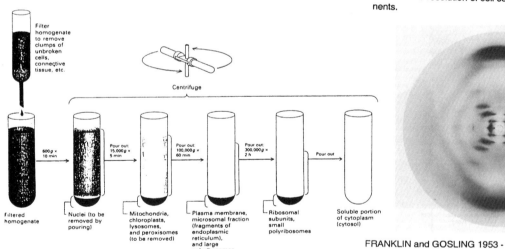

LODISH et al. 1995 - Ultracentrifugation resulting in separation of organelles and of macromolecules.

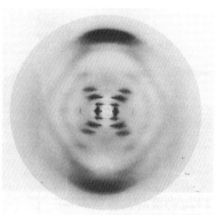

FRANKLIN and GOSLING 1953 - X-ray diffraction photograph of DNA. It was in this photograph that WATSON and CRICK found the clues to their double helical structure. The helical form is indicated by the dark areas that form a cross pattern in the center of the picture.

THE TECHNOLOGY THAT ALLOWED THE STUDY OF THE CHROMOSOME -- 6
1900 TO 2001.

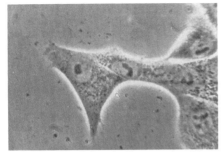

BRACHET 1957 - Cultured cell seen under phase contrast.

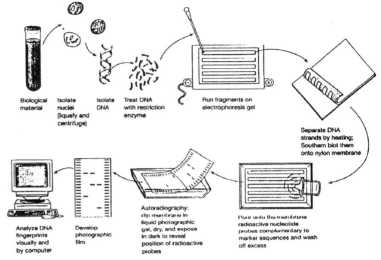

ARMS and CAMP 1995 - The DNA "fingerprinting" technique developed by SOUTHERN.

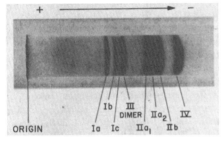

FAMBROUGH 1969 - Identification of histone proteins.

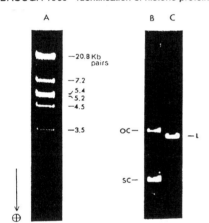

OLD and PRIMROSE 1980 - Separation of DNA segments after size.

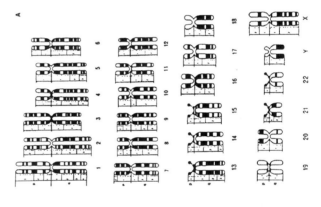

SANDBERG 1980 - Banding of the human karyotype.

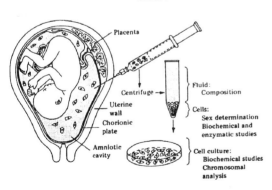

FRIEDMAN 1971 - Chromosome diagnostic during pregnancy.

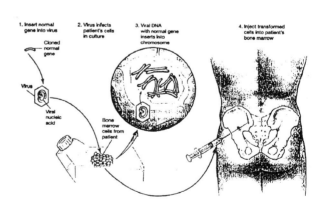

ARMS and CAMP 1995 - Injection of transformed cells into patients.

THE TECHNOLOGY THAT ALLOWED THE STUDY OF THE CHROMOSOME -- 7
1900 TO 2001.

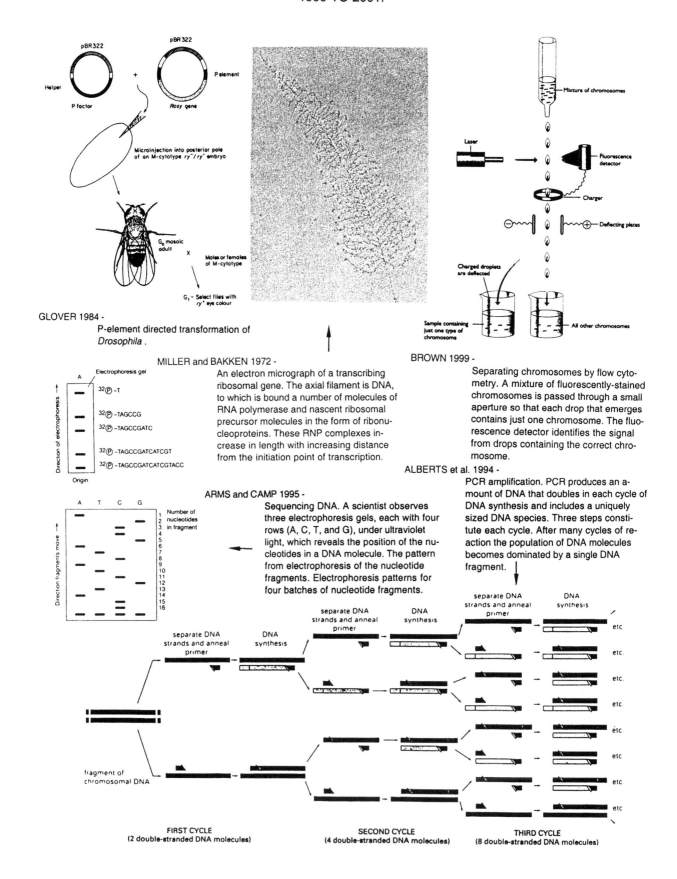

GLOVER 1984 -
P-element directed transformation of *Drosophila*.

MILLER and BAKKEN 1972 -
An electron micrograph of a transcribing ribosomal gene. The axial filament is DNA, to which is bound a number of molecules of RNA polymerase and nascent ribosomal precursor molecules in the form of ribonucleoproteins. These RNP complexes increase in length with increasing distance from the initiation point of transcription.

BROWN 1999 -
Separating chromosomes by flow cytometry. A mixture of fluorescently-stained chromosomes is passed through a small aperture so that each drop that emerges contains just one chromosome. The fluorescence detector identifies the signal from drops containing the correct chromosome.

ARMS and CAMP 1995 -
Sequencing DNA. A scientist observes three electrophoresis gels, each with four rows (A, C, T, and G), under ultraviolet light, which reveals the position of the nucleotides in a DNA molecule. The pattern from electrophoresis of the nucleotide fragments. Electrophoresis patterns for four batches of nucleotide fragments.

ALBERTS et al. 1994 -
PCR amplification. PCR produces amount of DNA that doubles in each cycle of DNA synthesis and includes a uniquely sized DNA species. Three steps constitute each cycle. After many cycles of reaction the population of DNA molecules becomes dominated by a single DNA fragment.

PART III

IN SEARCH OF THE EUKARYOTIC CHROMOSOME

Main stages in the discovery of the cell's structure and function

The study of the cell is an impressive example of the progress of science in the last 300 years. Thus, it may be appropriate to start with a short "cavalcade" that reviews the many steps connected with the discovery of the cell's organization.

A thin section of cork revealed to R. Hooke, in 1665, that this material had a structure that consisted of regular compartments limited by thin walls. More than one hundred years later the compartments were found to be filled with a fluid and to have a central body, the nucleus. By 1839 M.J. Schleiden and T. Schwann established the unity of organization of the plant and animal cells (T. Schwann 1847). During the beginning of the 20th century the nucleus was found to be surrounded by a series of organelles which included the chloroplasts and the mitochondria, but they lacked structural details. Inside the nucleus there was a smaller body, with constant shape, that was called the nucleolus (E.B. Wilson 1925). It was only with the use of the electron microscope that the endoplasmic reticulum, the centrioles, the plasma membrane, the nuclear envelope, the mitochondria and the chloroplasts became well defined structures with internal organization (S.W. Hurry 1967, D. Fawcett 1971). Twenty years later the peroxisomes, the Golgi apparatus, the ribosomes, the endosomes and the cytoskeleton were added (C. de Duve 1984). All these organelles were supposed to bathe in a chaotic soup agitated by Brownian movement. It is only in the last few years that most molecules in the cell, were found to have as part of their structure, an address tag, that marks their destination rigorously (A.P. Pugsley 1989). The cell now became represented three-dimensionally and new structures were added such as: basal bodies, lipid droplets, glycogen granules, transport and endocytic vesicles, microtubules and microfilaments (K. Arms and P.S. Camp 1995). The key features of bacteria and eukaryotic cells were compared by T. Cavalier-Smith (1991).

As we approached the turn of the new century, specific proteins, building proteasomes, were included into the cell's landscape (S. Patel and M. Latterich 1998) and two years later the relationship between protein subcellular localization and gene expression was investigated (A. Drawid et al. 2000). As the study of cancer reached the molecular level it was found that in healthy cells a protein called RAS will make a cell divide only if a receptor receives a molecular message telling it to do so, but the mutated RAS protein has a faulty shape and is permanently active (P. Jones 2001). The internalization of molecules into the cell, i.e. endocytosis, has turned out to be a complex machinery composed of many pathways (A. Sorkin 2000).

The different research periods illustrate the abyss existing between Hooke's empty cells and the successive populating of the cell's interior with a multiplicity of organelles and molecular processes that fill the interior of today's cells.

THE DISCOVERY OF THE CELL STRUCTURE AND FUNCTION -- 1

HOOKE 1665 - Cell walls found without contents.

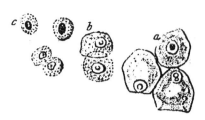

SCHWANN 1847 - Cells turned out to have nuclei and nucleoli.

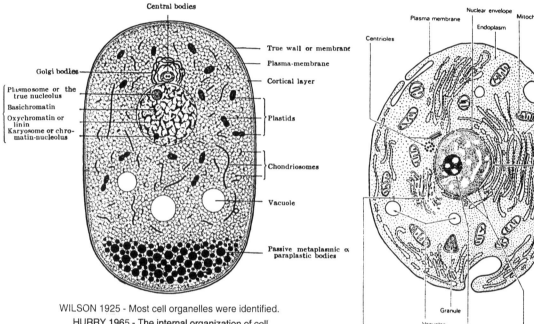

WILSON 1925 - Most cell organelles were identified.
HURRY 1965 - The internal organization of cell components was established.

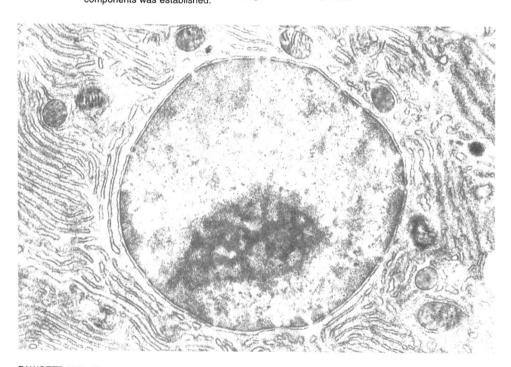

FAWCETT 1971 - Electron micrograph of a cell from the pancreas of a bat. The nucleus is in the center and is surrounded by the double-layered nuclear membrane. The smaller, round, striated structures in the cytoplasm are mitochondria. The long, thin structures are the endoplasmic reticulum.

THE DISCOVERY OF THE CELL STRUCTURE AND FUNCTION -- 2

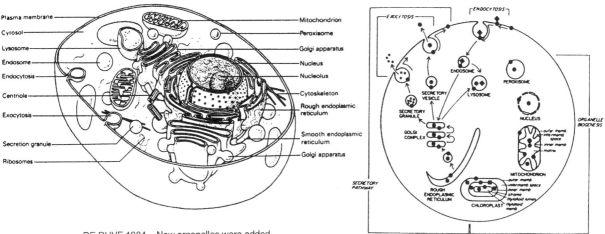

DE DUVE 1984 - New organelles were added.

PUGSLEY 1989 - Proteins were targeted to specific cell sites.

Key features of the four major sorts of cell

	Bacteria		Eukaryotes	
	Eubacteria	Archaebacteria	Archezoa	Metakaryota
DNA				
Shape	Circular	Circular	Linear	Linear
Attached to	Plasma membrane	Plasma membrane	Nuclear lamina	Nuclear lamina
Introns in genes				
5'OH protein-spliced	−	tRNA, rRNA	?	tRNA
3'OH self-spliced	tRNA	−	?	rRNA
(* mitochondria and				mRNA*
chloroplasts only)				tRNA*
3'OH spliceosome-spliced	−	−	?	mRNA
Nuclear pore complex	−	−	+	+
Histones and nucleosomes	−	−	?	±
Replicons per chromosome	1	?	?	many
Capped and tailed mRNA	−	−	?	+
Endomembrane system				
(RER, lysosomes)	−	−	+	+
Golgi dictyosomes	−	−	−	+
Cytoskeleton	−	−	+	+
Cytosis (exo-, endo-, er-)	−	−	+	+
Nucleus and mitosis	−	−	+	+
Extrusomes	−	−	−	±
Extracellular proton-driven				
flagella	±	±	−	−
9 + 2 cilia / centrioles	−	−	±	±
Ribosomes	70 S	70 S	70 S	80 S
Polygenic mRNA	±	±	?	−
Isoprenoidal ether lipids	−	+	−	−
Acyl ester lipids	+	−	+	+
Mitochondria	−	−	−	+
Peroxisomes	−	−	−	+
Chloroplasts	−	−	−	±
Murein walls	+	−	−	−
N-Asparagine-linked				
glycoproteins	−	+	+	+
Cellular endosymbionts	−	−	±	±
Sex (syngamy and meiosis)	−	−	±	±
AAA lysine biosynthetic				
pathway	−	−	?	+
Sterols	−	−	?	+
3 separate RNA polymerases	−	−	?	+
Ca^{2+}/calmodulin/inositol				
phosphate messenger system	−	−	?	+

CAVALIER-SMITH 1991 - Bacteria and eukaryotic cells were compared.

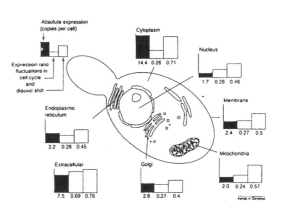

DRAWID ET AL. 2000 - Expression patterns were described in major compartments of the yeast cell.

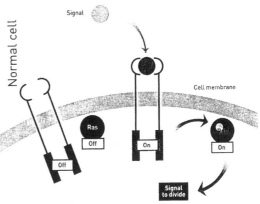

JONES 2001 - In healthy cells, a protein called Ras will make a cell divide only if a receptor receives a molecular message telling it to do so.

THE DISCOVERY OF THE CELL STRUCTURE AND FUNCTION -- 3

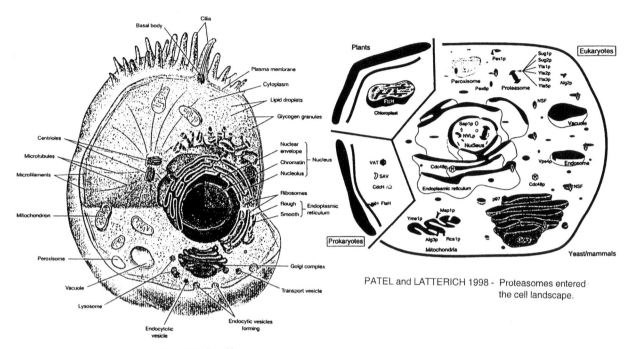

ARMS and CAMP 1995 - The cell displayed in a three-dimensional view.

PATEL and LATTERICH 1998 - Proteasomes entered the cell landscape.

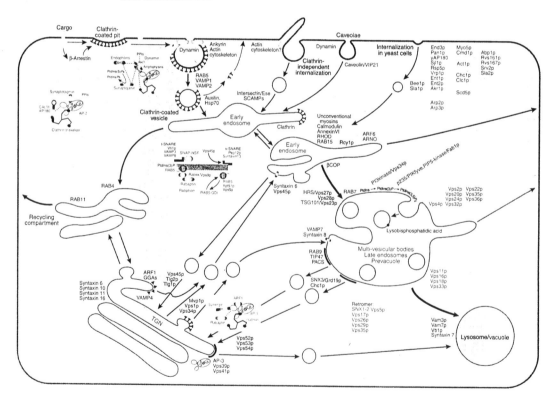

SORKIN 2000 - The endocytosis machinery of a cell. The figure shows the major proteins involved in the regulation of endocytic pathways including mammalian and yeast proteins.

The nucleus versus the cytoplasm. Which was most important?

Was it the nucleus or the cytoplasm that contained the critical information that produced a new cell or a new organism with its characteristic properties? The question was topical already in the 19th century and it led E.G. Balbiani (1885) to cut into three pieces the protozoa *Stentor*. Each of the pieces contained several fragments of the animal's multiple nucleus. Within 24 hours each fragment had regenerated a complete animal. Soon after that M. Verworn (1888) attempted a more exact experiment. The protozoa *Stylonychia* was cut into three pieces. The middle piece containing the two nuclei regenerated a perfect animal whereas the upper and lower parts, without nuclei, quickly perished. J. Hämmerling (1953) took up this question by experimenting with the unicellular alga *Acetabularia*. He grafted *A. mediterranea* on *A. crenulata*, two species with quite different body forms. It turned out that the type of nucleus present in the graft was the decisive factor in shaping the alga's phenotype. The nucleus dominated over the cytoplasm. But what kind of morphogenetic substances carried the message from the nucleus to the final organism pattern? J. Hämmerling already noticed clear-cut differences in nuclear action depending on the distance from the nucleus but the molecules involved remained elusive (J. Brachet 1974).

Cell differentiation during embryonic development was an area in which genetics had been retarded. For decades all interest was concentrated on progeny analysis and on gene identification. By the 1970s this trend started to change. J.B. Gurdon (1974) carried out a series of nuclear transplantations in which he introduced the nucleus from an intestine cell, or from a skin cell, of a *Xenopus* tadpole, into a recipient egg that lacked nucleus (following ultraviolet irradiation). This egg developed producing a tadpole or a normal adult animal (J.B. Gurdon 1974). It may be recalled that the technique of transplanting blastula cell nuclei into enucleated eggs was originally developed by R. Briggs and T.J. King twenty years earlier (1953). B. Mintz (1971) developed another technique in which she fused, *in vitro*, cleavage stage embryos from mice, carrying genes for dark and for light cell pigmentation. The resulting embryonic chimaeras were implanted into the uterus of a female mouse. This gave birth to normal mice having their fur displaying large bands of light and dark pigment. These studies later led to experiments in which the fate of cancer cells was followed during mammalian development.

At the same time as the nucleus became accepted as the source of genetic information carried out by its chromosomes, it became evident that the cytoplasm was not wholly devoid of important genetic information. As early as 1937 C. Correns described cases of what was called maternal or extrachromosomal gene inheritance. Crosses in plants showed that, besides the nucleus, the chloroplasts carried genetic information of their own. In animals the inheritance of body pattern turned out to be dependent on that of the mother.

One had to wait until the 1960s for the demonstration of DNA in chloroplasts and in mitochondria, by combining electron microscopy with tritium autoradiography. This led to a biochemical and molecular analysis of these organelles which resulted in the location of genes in their DNAs. Subsequently the DNAs of the mitochondria of many species have been sequenced and the information has been used to establish possible phylogenetic relationships between higher vertebrates including humans. The organization of human and yeast mitochondrial DNAs can be compared by putting together the results of P. Borst and L.A. Grivell (1981), of A. Chomyn et al. (1986) and of other research groups. By using specific fluorescent dies the DNA of the mitochondria can be visualized in the cell of *Euglena gracilis* (Y. Huyashi and K. Veda 1989). A list of the genes encoded by the chloroplast DNA of a liverwort, has been assembled by K. Ohyama et al. (1986), showing the many DNA sequences that code for proteins essential for photosynthesis. The regulation of mitochondrial integrity and of its dysfunction is being actively investigated at present (E.H.-Y.A. Cheng 2001).

How has this interplay between the genetic information of the cytoplasm and that of the nucleus become established and under which molecular rules did it continue to evolve? At present there are no answers to these questions.

THE NUCLEUS VERSUS THE CYTOPLASM. WHICH WAS MOST IMPORTANT?

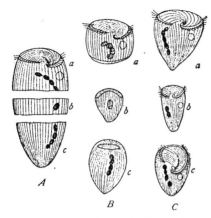

BALBIANI 1885 - Pieces with fragments of nucleus regenerated.

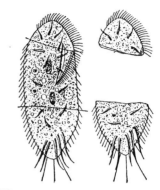

VERWORN 1888 - Pieces without nucleus perished.

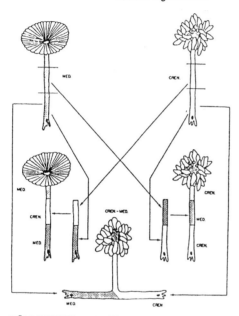

HÄMMERLING 1953 - The nucleus present in the graft determined body pattern.

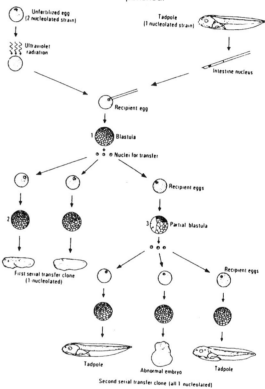

GURDON 1974 - Transplantation of intestine cell nucleus into enucleated egg resulted in tadpole.

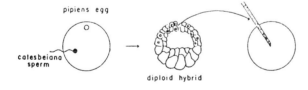

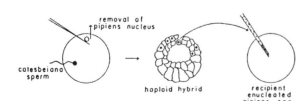

BRIGGS and KING 1953 - Blastula nuclei were transplanted into enucleated frog eggs.

GURDON 1974 - Normal adult of *Xenopus* obtained from transplantation.

The description of cell division
An impressive transformation was accompanied by directed cellular movements

The division of cells into two daughter units was accompanied by the sudden emergence of individualized round-shaped bodies that stained strongly with most dyes. These bodies started by assembling in the middle of the cell and astonishingly they were able, by themselves, to move to two opposite sides, which were later called poles. Even more astonishing was that their movement revealed an impressive order in the distribution of these elements into compartments that became the two daughter cells.

The same year as the major capitals of Europe, were shaken by revolutions arising from acute conflicts, derived from the rapid industrial expansion, the division of the cell was described in detail by W. Hofmeister (1848). From the beginning the division of the cell was studied mainly in somatic tissues or on dividing eggs. In these tissues the type of division which occurred led to the maintenance of the same chromosome number in the ensuing daughter cells. This situation was studied in detail by W. Flemming (1882) and by E. van Beneden and J. Julin (1884). The chromosomes exhibited a singular behaviour: they duplicated their mass and they regularly distributed their products to the spindle poles. The daughter chromosomes were easily identified as they appeared as darkly stained bodies quite distinct from the surrounding cytoplasm. By 1896 M. Heidenhain described another singular property. This was the emergence of the chromosomes in the apparently empty interphase nucleus. They started to show up in the nucleus as nearly indistinguishable thin threads, to successively condense into short and thick rods. By 1925, when E.B. Wilson published his classic work "The Cell in Development and Heredity", the different stages of mitosis: prophase, metaphase, anaphase and telophase, were well recognized and found to be, not only comparable, but essentially identical in invertebrates, such as the sea urchin, plants, such as the onion and vertebrates such as the salamander (E.B. Wilson 1925).

The cell division was a process loaded with other surprising events. The spindle that became most conspicuous at metaphase disappeared by telophase, and the same thing happened to the chromosomes when they became imprisoned inside the new nuclear membrane. What mechanisms made possible this behaviour of permanent change combined with permanent order? One was confronted with an antithetical display which was an integral part of the cell's construction.

THE DESCRIPTION OF CELL DIVISION.
AN IMPRESSIVE TRANSFORMATION WAS ACCOMPANIED BY DIRECTED CELLULAR MOVEMENTS.

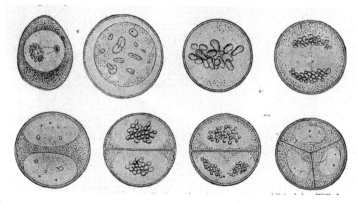

HOFMEISTER 1848 - First description of the division of the cell.

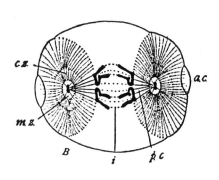

VAN BENEDEN and JULIN 1884 - The chromosomes moved on the spindle.

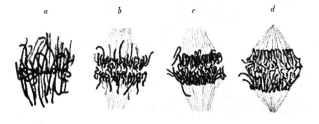

FLEMING 1882 - Many stained bodies present during cell division.

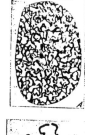
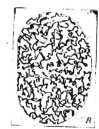

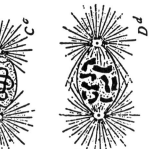

WILSON 1925 - The different stages of mitosis.

HEIDENHAIN 1896 - The condensation of the chromosomes.

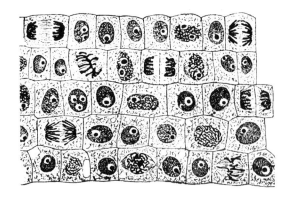

WILSON 1925 - Mitosis in the onion.

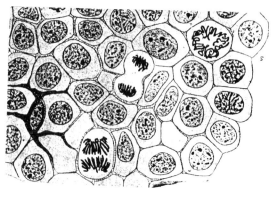

WILSON 1925 - Mitosis in the salamander.

Meiosis was another unexpected property
The cell could reduce its chromosome number

Few processes consist of so many unique molecular events as the division by which a cell reduces its chromosome number to half.

As the study of the cell division extended from somatic tissues to the sexual organs of plants and animals, cytologists were confronted with another variant of this process, which was later called meiosis (from the Greek word "meion" that means smaller) since it represented a reduction of the chromosome number from diploid to haploid in germ cells.

The novel and central feature was that the cell divided, but the chromosomes did not accompany the cytoplasm in this process, they simply "refused" to divide as they did in somatic tissues, *i.e.* the cell doubled but the chromosomes did not. The germ cells got half of the chromosome number, the diploid number being reestablished after fertilization. Like mitosis, meiosis turned out to be alike in plants and animals. But many unique phenomena were found to accompany this division.

Meiosis became characterized by the following series of programmed events.
1. A long DNA synthesis at interphase, which also extended into prophase in certain species. This last synthesis occurred in connection with ribosomal DNA amplification and other phenomena.
2. A very long prophase.
3. Formation of distinct chromomeres at early and middle prophase I.
4. Pairing of homologous chromosomes.
5. Appearance of DNA components characteristic of certain stages of prophase I.
6. Appearance of a synaptonemal complex between paired chromosomes.
7. General occurrence of crossing over.
8. Formation of chiasmata.
9. Packaging of chromosomes from prophase to metaphase.
10. Separation of the chromosomes of each bivalent to opposite spindle poles.
11. No DNA synthesis at interphase II.
12. Wide separation of the two chromatids of each chromosome in "X form" at prophase II.
13. Reduction of the chromosome number to the haploid condition.
14. Appearance of novel chromosome combinations in the products of this division.
15. Formation of informational RNAs and proteins.

The duration of the cell cycle was affected by internal as well as external factors. Among the internal factors are the genetic constitution and the stage of organism development; external factors which affect the cycle most conspicuously are temperature and light conditions (D.L. Lindsley and E. Lifschytz 1971). Mitosis may take only 5 min, as in *Xenopus* embryos, but may last for 13 hours, as in HeLa cells at 33 °C. Meiosis in *Triticum aestivum* may have a duration of 24 h and in *Tradescantia paludosa* of 126h (A.F. Dyer 1979).

The most remarkable events were: the absence of DNA synthesis at the interphase between the first and the second divisions, the occurrence of crossing over, the appearance of a synaptonemal complex and the display of distinct chromomeres along the chromosomes.

Among the first discoverers of meiosis were W. Flemming (1887), F. Meves (1896) and L. Guignard (1898). By this time A. Weismann (1893) was already considering the possibility of recombination among chromosome segments, which later became known as crossing over, but chromosomes did not pair end to end as he had envisaged. As the new century was born W.S. Sutton (1902) described in a grasshopper the exact formation of paired chromosomes at meiosis. The events occurring during the long meiotic prophase, as well as the designation of the different stages into which it was divided, were established by L.W. Sharp (1934) and C.D. Darlington (1937). The synaptonemal complex that forms between paired chromosomes was identified in the electron microscope by M.J. Moses in 1956 and 1968. Its more detailed pattern was revealed by M. Westergaard and D. von Wettstein (1970). Another particular feature is the lampbrush structure that the Y chromosome of *D. hydei* assumes during meiosis (O. Hess 1971).

Crossing over is now starting to be explained in molecular terms and the proteins that direct DNA exchanges are being isolated. This phenomenon is still sometimes described as a random process, but is now turning out to be ordered by rigid protein-DNA interactions. The repair of double strand breaks (DSB) during DNA recombination and the specific proteins participating in it have been studied by J.E. Haber (1999).

Meiosis still remains an unchartered area exposing our great ignorance of chromosome behaviour. Since the discovery of G. Mendel's rules of inheritance it is possible to predict the progeny that arises following a given cross. But this prediction is limited to statistical information. It is not possible to predict in advance the genetic constitution of a single individual.

Chromosomes obey genetic mechanisms that

in some species rigorously oblige homologous chromosomes to separate orderly. Confirmation of this situation was obtained in insect species, such as *Sciara*, in which the chromosomes of the mother moved to one pole, whereas those from the father moved to the opposite pole (H.V. Crouse 1965). This phenomenon also occurs in plants like *Rosa canina* (C.D. Darlington 1937). In *Drosophila* segregation "distorter genes" have been localized which change a 50:50 segregation into a 99:1 chromosome segregation. One such gene is located in chromosome II of *D. melanogaster* (J.F. Crow 1979).

The recent finding of what has been called "the pachytene checkpoint" supports these earlier results. The term checkpoint refers to the control mechanisms that enforce order of cell-cycle events during meiosis. If chromosome pairing and recombination have not been completed the cell is arrested at the prophase of meiosis (pachytene). This control mechanism prevents chromosome missegregation that would lead to the production of abnormal gametes (G.S. Roeder and J.M. Bailis 2000).

Hence the chromosomes contain genes that direct their assortment. Once these genes are fully mapped, and their products isolated in the test tube, the outcome of meiosis may be predicted more accurately and the determination of the genetic constitution of single individuals will become easier to control.

MEIOSIS WAS ANOTHER UNEXPECTED PROPERTY.
THE CELL COULD REDUCE ITS CHROMOSOME NUMBER -- 1

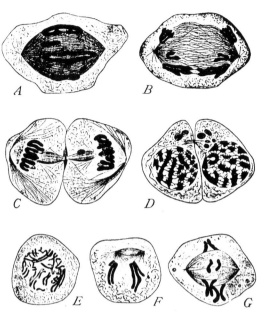

FLEMMING 1887 - Some stages of meiosis in the
and MEVES 1896 salamander.

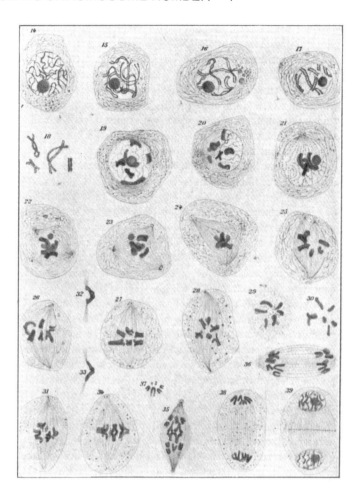

GUIGNARD 1898 - The correct sequence of meiosis.

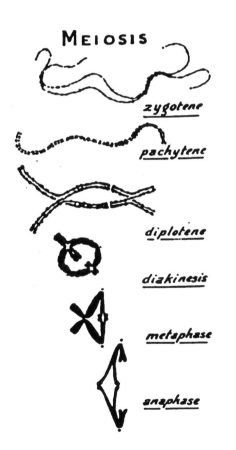

DARLINGTON 1937 - Understanding of the pairing process.

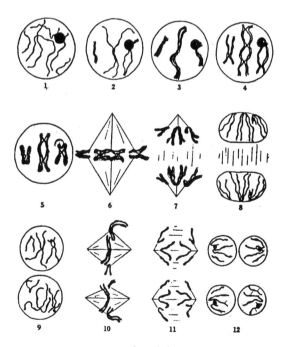

SHARP 1934 - Detailed description of meiosis.

MEIOSIS WAS ANOTHER UNEXPECTED PROPERTY. THE CELL COULD REDUCE ITS CHROMOSOME NUMBER -- 2

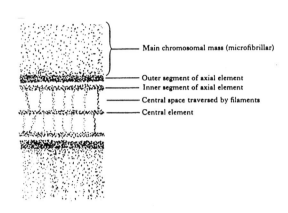

MOSES 1968 - The components of the synaptonemal complex.

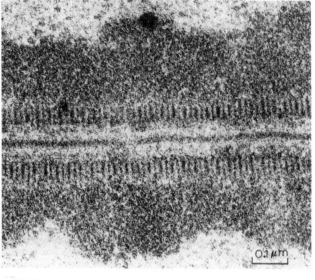

WESTERGAARD and WETTSTEIN 1970 - The pachytene picture of the complex in the E.M.

HESS 1971 - The Y chromosome of *D. hydei* at the lampbrush stage in primary spermatocytes.

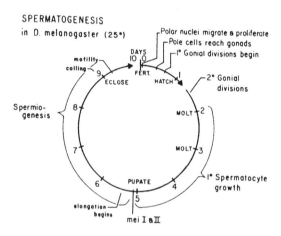

LINDSLEY and LIFSCHYTZ 1971 - The male germinal cycle in relation to the life cycle of *D. melanogaster*.

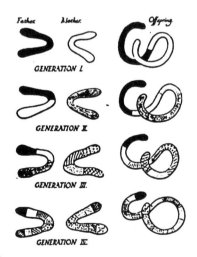

WEISMANN 1893 - Recombination among chromosomes.

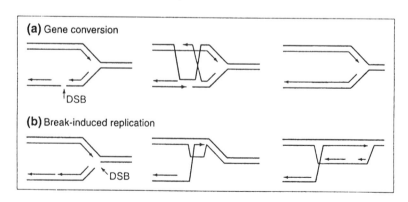

HABER 1999 - Repair of double strand breaks (DSB) during DNA replication.

The maintenance of identity of the chromosome during interphase was accompanied by constancy and variability of pattern in different tissues

The nucleus appeared in most cells as an empty structure devoid of chromosomes. What happened actually to the chromosomes that were so distinct during cell division but which disappeared during the "resting stage"? Were they amorphous structures that dissolved and lost their shape and continuity, or did the passage through the resting stage only represent a period in which they modified their outer appearance?

The permanence of the chromosomes, during what was later called "interphase", and the maintenance of their identity during this stage, were confirmed by observations carried out by T. Boveri and by K. Belar in the early decades of the 1900s. This was one of the key results that led to the establishment of the chromosome theory of heredity. The arrangement of the chromosomes, seen at the beginning of a division, was found to correspond exactly to that observed in the previous division. Moreover, when chromosomes contained heterochromatic segments, or a whole chromosome was heterochromatic, as in the case of large sex chromosomes, these could be recognized all the way from telophase throughout interphase to the following metaphase (V. Gregoire 1932, F.W. Tinney 1935).

Later arose another equally pertinent question. Was the detailed pattern of a chromosome, observed during the early stages of division, a constant feature or did it change from cell to cell or among individuals from the same species? The pachytene stage of the prophase of meiosis, and the salivary glands of Diptera, turned out to display a chromosome pattern consisting of chromomeres or "bands", which varied in size and configuration along the chromosomes. The detailed analysis of the chromomere pattern at pachytene, with its groups of large chromomeres called "knobs", revealed that the pattern was repeated with high fidelity from cell to cell both in animals (D.H. Wenrich 1916) and plants (A.E. Longley 1939). The differences in the shape of the bands found along the chromosomes were accurately drawn for every chromosome and they were numbered accordingly. The configuration of individual bands changed during the building of "puffs", as the development of the insect proceeded, but each individual band could still be recognized with great accuracy at the different stages, the basic chromosome pattern being maintained independently of the specific activity of individual bands (H.J. Becker 1959, H.D. Berendes 1965).

The molecular mechanisms which are responsible for the emergence of the chromomere and band patterns, as well as those that direct their reappearance at a specific stage of the organism's development, remain obscure.

What has confused us throughout so many years of chromosome research is that, at the same time as the chromosome is a structure that displays an extreme permanence of pattern and of organization, it also shows an extreme variability. How is this possible since it has behaved in this way for over 2 billion years? The antithetic nature of the chromosome is in a way not so remarkable, since molecules, cells and organisms display these two opposite properties. However, in the chromosome this situation seems to have attained rather extreme dimensions. A chromosome may change its appearance so drastically that it is difficult to believe that it was the same structure that was identified a few hours earlier at a different stage of the same cell division or at another cell event occurring within the same organism.

Chromosome I of the plant *Ornithogalum virens* measures 110 microns at pachytene but it is only 6 microns long at the metaphase of mitosis. The fine chromomere structure, that it displays at pachytene, disappears completely at metaphase of mitosis, the chromosome becoming a solid mass. Hence the difference in length, shape and appearance is so drastic that no one would believe that this was the same structure studied only a few stages apart. The same is true for the other two chromosomes of this species (A. Lima-de-Faria et al. 1959). In animals, in which the comparison is made between different meiotic stages, the same situation arises. In the cricket *Acheta* the behaviour of the X chromosome is equally dramatic. The X is a distended structure that has 41.7 chromomeres in oocytes but none in spermatocytes, where it shows up as a compact dark stained body. Moreover, in the spermatogonia it occurs as a regular mitotic chromosome (A. Lima-de-Faria 1973).

During the so-called "lampbrush stage", found in the diplotene of several animal species, the chromomere itself may change its pattern in a most unexpected way. From building a small spiral in the chromosome fiber it may balloon in the form of a long loop, to become a highly differentiated string in which molecular reactions occur (O. Hess 1971).

Hence, the chromosome is one of the most accomplished magicians. It can transform its shape, in such a way, that it would have been impossible to relate it to its previous structure, if there was no knowledge of its material continuity throughout cell development.

THE MAINTENANCE OF IDENTITY OF THE CHROMOSOME DURING INTERPHASE WAS ACCOMPANIED BY CONSTANCY AND VARIABILITY OF PATTERN IN DIFFERENT TISSUES.

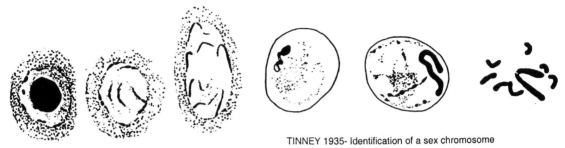

GREGOIRE 1932 - Permanence of chromosomes during interphase.

TINNEY 1935 - Identification of a sex chromosome from interphase to metaphase.

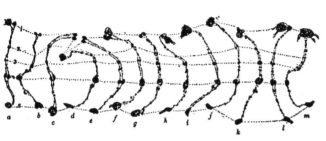

WENRICH 1916 - Repetition of the chromomere pattern from cell to cell in animals.

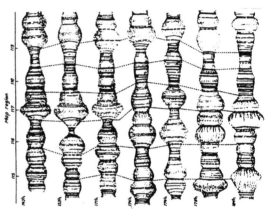

BERENDES 1965 - Maintenance of band pattern in *D. hydei*.

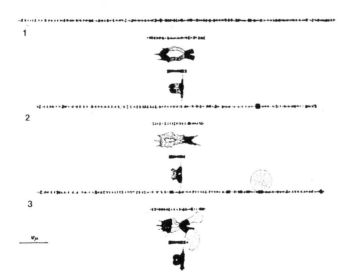

LIMA-DE-FARIA et al. 1959 - Chromosomes 1, 2 and 3 of *O. virens* at five different stages: pachytene, pollen mitosis, diakinesis, root mitosis and metaphase I of meiosis.

HESS 1971 - Extreme variation in the chromomere pattern of lampbrush chromosomes (a, b, c, d).

PART IV

THE THREE UNIQUE REGIONS OF THE EUKARYOTIC CHROMOSOME

The centromere
A Pandora's box of unearthed properties

A chromosome is a structure which contains an organization not found elsewhere in any other cell organelle. The centriole, the Golgi apparatus, the mitochondrion, the chloroplast, the endoplasmic reticulum, the nuclear membrane, none of them has properties comparable to those of that key element – the chromosome.

Three unique features define the chromosome: it has a centromere, a telomere and a nucleolus organizer. Each of these regions has a specific function. The centromere allows the chromosomes to move to the spindle poles by becoming functionally coupled to protein fibers. The telomere closes the terminal region of each chromosome arm, blocking a further extension of its length. The nucleolus organizer produces a body, consisting mainly of ribosomal RNA and protein, which remains attached to the chromosome body at this segment. This region, is found in only one or a few chromosomes of the complement, but its presence is obligatory.

Besides these main properties, each of the three regions, has secondary functions which are equally important. The centromere, is the organ which controls the distribution of several properties such as: crossing over, chiasmata, heterochromatin, chromomere gradients and others (C.D. Darlington 1937, E.B. Lewis 1950, J. Schultz and H. Redfield 1951). The telomere, likewise, interferes with the occurrence of a series of properties in its vicinity (L.H.T. Van der Ploeg 1990, D. De Bruin et al. 2000). The nucleolus organizer also affects the occurrence of other chromosome features such as the chromomere pattern (A. Lima-de-Faria 1956). Moreover, these three specialized regions have one feature in common – they inhibit the presence of identical regions in their close vicinity. Centromeres that happen to be moved close to other centromeres are silenced. This has been demonstrated by translocations both in wheat and in humans (E.R. Sears and A. Camara 1952, E. Niebuhr and F. Skovby 1977). Telomeres are normally not accompanied by other telomeres having interstitial positions. In the case of the nucleolus organizer its inhibitory action is so strong, that among over 700 species in which the location of the nucleoli has been determined, there is not a single case in which two nucleolus organizers occur within the same chromosome arm. Nucleolus organizers, occurring in several chromosomes of the same complement, are known. That is for instance the case in humans who have 5 nucleolar chromosomes. However the most common situation is a genome with a single nucleolus organizer. Remarkable is that among these same 700 species, there is a single case in which one chromosome possesses two nucleolus organizers but located in different arms. Hence, the inhibitory effect of the organizer is apparently so strong that it has not allowed the establishment of another identical region within the same arm and only, in exceptional circumstances, has permitted the establishment of a similar one on the opposite arm of the same chromosome.

These inhibitory effects which extend along the chromosome arms, have been considered to be a primary mechanism responsible for the maintenance of chromosome unity (A. Lima-de-Faria et al. 1984, 1991) (see pages 34 and 176).

However, exceptional situations occur. In the case of centromeres certain plants (*Luzula*) and animals (Coccids) possess chromosomes with, so-called, diffuse centromeres in which many such regions are located along the entire chromosome body (N. Malheiros et al. 1947, S. Hughes-Schrader 1942). These are isolated cases.

Telomeres may also happen to have interstitial positions. Chromosomes, in rye and in maize, can behave normally when large terminal segments are deleted. Other internal regions take over the function of sealing the chromosome arm. Interstitial telomeres also occur in lizard chromosomes, as detected by *in situ* hybridization with telomeric sequences $TTAGGG_n$ (C.E.V. Bertolotto et al. 2001). But these cases do not represent the rule.

The molecular mechanisms that direct these inhibitory effects, and those that interfere with their action, are a fertile area of future research, for only when they are elucidated, will we start to understand how the chromosome is organized as a unique functional unit distinct from all other cell organelles.

The centromere has been the source of so much study that a whole book has recently been written dealing with its molecular organization (K.H.A. Choo 1997) but so much is known about it that a series of volumes would be necessary to cover the knowledge accumulated on this chromosome region.

Let us summarize the most relevant data on the following items: 1) origin, 2) structure and DNA content, 3) location, 4) cycle of division, 5) movement on the spindle, 6) molecular organization, 7) occurrence of iso-chromosomes, 8) function of dicentrics and polycentrics, 9) unanswered questions.

Origin – In 1932 M. Navashin refuted the idea that new chromosome numbers could arise by simple fragmentation or fusion of chromosomes, independently of the number of centromeres present in the chromosome set. He clearly developed the idea that only chromosomes possessing one single centromere could perpetuate themselves through a considerable number of cell generations and that polycentric chromosomes could not do so. Secondly, he stated that centromeres do not arise *de novo* and do not lose their properties in the course of evolution. He enunciated these two statements in the following way: "the chromosome number is conditioned by the number of functional kinetic bodies, the latter not being subject to *de novo* formation". M. Navashin's idea has been of great value, inasmuch as it called the attention of his colleagues to the important role played by the centromere in chromosome movements. The validity of the first statement of M. Navashin, as regards the chromosomes possessing a localized centromere, was soon confirmed. Polycentric chromosomes represent exceptional cases. The second statement, *viz.* that centromeres do not arise *de novo*, has been generally accepted owing to the fact that evidence to the contrary was not found for many years. At present this question is under debate.

Structure and DNA content – The centromere was considered to be structureless and lacking DNA for many decades. F. Schrader (1944) summarized this situation by putting together, several hypothetic interpretations of the centromere structure, that had been proposed until then. This situation changed when a well defined pattern consisting of chromomeres, identical to those found in the chromosome arms, was described in the pachytene chromosomes of rye (A. Lima-de-Faria 1949). The structure of the centromere was divided into three distinct zones: The exterior zone consisting of almost unstained fibrillae. A second zone composed of two pairs of chromomeres. A third interior zone formed by deeply stained fibrillae which later turned out to consist of minor chromomeres. It became evident that this pattern was symmetrical, a plane passing through the middle of the centromere dividing it into two parts related one to the other as an object to its image in a plane mirror. Such a finding allowed explaining the occurrence of functional iso-chromosomes (chromosomes with two identical arms) which before could not be understood and which resulted from a division occurring in the middle of the centromere.

Already in the 1950s it was possible to apply the Feulgen test to control whether a region possessed DNA. This cytochemical procedure revealed that DNA was also present in the centromeric chromomeres (A. Lima-de-Faria 1950). But at this time this finding had little genetic significance, since it was generally believed that the gene consisted of protein in higher organisms. It took three years before the J.D. Watson and F.H.C. Crick model was proposed, but at this time there was still no reason to believe that there were genes at the centromere.

Location – The location of centromeres would seem to have been a simple matter, but it turned out to be a source of confusion. Chromosomes with median and subterminal centromeres (called metacentrics and sub-metacentrics respectively) were easily identified by the formation of a chromosome constriction, without structure, usually called a "gap". But when the centromere was located terminally (telocentrics) several authors denied their existence, stating that there was a very small arm that could hardly be seen. This confusion disappeared when the centromere was found to have a specific structure, which could be identified at the terminal region. This finding was confirmed in plants and mammals including mice. The different types of location were summarized by G.L. Stebbins (1971).

Cycle of division – Here we enter another area where discordance has reigned. Since C.D. Darlington in the 1930s put forward the idea that the centromere was the last region of the chromosome to divide, at the onset of anaphase, this statement has been repeated in most textbooks and papers. However, the improved cytological techniques of the 1950s and the electron microscopy of the 1980s could show clearly that the centromere is one of the last regions to divide but not the last to hold the chromatids together. This is accomplished by the proximal regions of the arms situated close to the centromere on its both sides (A. Lima-de-Faria 1955, V.J. Goyanes and J.B. Schvartzman 1981). It is true, that in some cases, such as in polytene chromosomes, the distinction between the two regions is not easy to establish. This has resulted in including the arms into what has been called the "centromeric region". This difficulty became even more acute when describing chromosome movements. The new results on the structure of the centromere and on its cycle of division were summarized in a diagram by K.R. Lewis and B. John (1963).

Movement on the spindle – From the beginning chromosome movements were considered to be directed by electrostatic forces (R.S. Lillie 1905, 1911). As late as the 1950s they were thought to be moved by "elastic forces". It was only the advent of molecular biology that permitted an exact study of this phenomenon.

There has been a geographic separation concerning the use of the terms centromere and kinetochore. Originally Europeans tended to use the first and Americans the second designation. At present both terms are being applied: the kinetochore is reserved for the zone that attaches to the spindle fibers and the centromere comprising the rest of the structure. This distinction derives mainly from the electron microscope work of B.R. Brinkley and E. Stubblefield (1970) and has later been depicted by H. Lodish et al. (1995). The microtubule connected with the spindle pole body is considered to attach at the kinetochore plates (or centromere core) by means of the centromere-binding factor (CBF), a complex of three proteins.

Molecular organization – The molecular organization of the centromere is not so well known as that of telomeres. Yeast centromeres are much smaller than those of mammals which consist of several millions of base pairs. This is not surprising since the yeast chromosomes themselves are 100 times smaller than those of higher eukaryotes. The centromeres of yeast and *Drosophila* are formed by three DNA segments, I, II and III, which have been sequenced by L. Clarke and J. Carbon (1985). Their spatial relationship to the microtubule of the spindle, and to the nucleosome core of the chromatid, was depicted by J. Carbon (1984). H. Lodish et al. (1995) marked the position of the protein complex consisting of three proteins, which functions as the centromere binding factor (CBF). The organization of the protein components of yeast centromeres, in relation to the microtubule, has been described in detail by A.F. Pluta et al. (1995). The corresponding situation in mammalian chromosomes has also been visualized by A.F. Pluta et al. (1995). Recently a partial DNA sequencing was obtained for the five centromeres of the plant *Arabidopsis* (The Arabidopsis Genome Initiative 2000).

Function of dicentrics and polycentrics – Physics had an impact on genetics that was both fruitful and negative. The positive aspect came from the introduction of technologies, such as electrophoresis. The negative consequence was that many genetic events tended to be interpreted in mechanical terms. When chromosomes with two centromeres were discovered, it was quickly concluded that they could not survive because of the mechanical stress between the two centromeres, that would lead to chromosome disruption. It has taken time to abandon this type of interpretation. At present many instances are known, from plant, animal and human chromosomes, that dicentrics can persist, and function quite well, depending on the number of DNA bases present between them (E. Niebuhr and F. Skovby 1977, B.A. Sullivan and H. Willard 1998). The mechanism of interaction between centromeres is starting to be unveiled. In polycentrics and in chromosomes with diffuse kinetochores there seems to be a comparable interaction that still remains obscure.

Unanswered questions – An impressive amount of questions can be asked concerning the function, delimitation and division cycle of the centromere.

Surprisingly, chromosomes can move both with and without spindle. In early prophase nuclei chromosomes move constantly in the nuclear sap enclosed by the nuclear envelope. The ends of chromosomes come together forming the so-called "bouquet stage". This is actually seen in the films made by A.S. Bajer using the cells of several plant species. The chromosomes are seen in permanent movement inside the nucleus. Hence, the movement that takes place later at anaphase, when the nuclear envelope disappears and the spindle forms, is a secondary process that has evolved subsequently. The centromere-spindle apparatus, with its many molecular components, is reminiscent of the complexity of the events in a ribosome at the moment of protein synthesis. We are dealing with phenomena of molecular recognition and of molecular self-assembly that will be a fertile area of future studies.

Much is known about DNA replication but that is not tantamount to chromosome replication or separation. This last process contains many intriguing components. No less than four different regions of the chromosome have distinct cycles of chromatid separation: the ends, the median regions of the arms, the proximal regions of the arms and the centromeres. Practically nothing is known of what causes this local disparity (K.H.A. Choo 1997).

The delimitation of the centromere is not an academic problem. The inclusion or exclusion of adjacent satellite DNAs into its structure, will only be decided when the function: 1) of the active region (kinetochore), 2) of the centromere itself and 3) of the surrounding DNA sequences, are better understood in molecular terms.

THE CENTROMERE - A PANDORA'S BOX OF UNEARTHED PROPERTIES -- 1

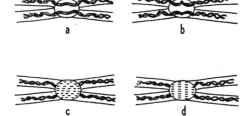

SCHRADER 1944 - Hypothetical structure of the centromere according to various authors. *a* kinosome (Matsuura 1941), *b* spindle spherule (Sharpe 1934), *c* oriented micellae (Darlington 1939), *d* arrangement of micellae leading to misdivision (Nebel 1939).

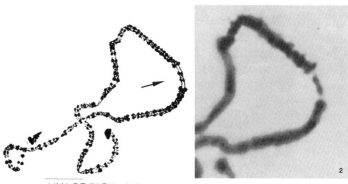

LIMA-DE-FARIA 1949 - Camera lucida drawing of a chromosome of rye at pachytene showing the centromere (arrow) consisting of three regions. An exterior zone of almost unstained fibrillae, a chromomeric zone composed of two pairs of centromeric chromomeres and the interior zone consisting of deeply stained fibrillae. To the right a photograph of the centromere and part of the arms.

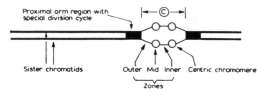

LEWIS and JOHN 1963 - Diagram summarizing LIMA-DE-FARIA's results (1949 and 1955) on the chromomere structure of the centromere with its three zones, and the special division cycle of the proximal arm regions.

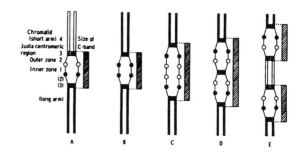

NIEBUHR and SKOVBY 1977 - Formation of dicentrics in human chromosomes based on the chromomere pattern of the centromere, following breakage and fusion.

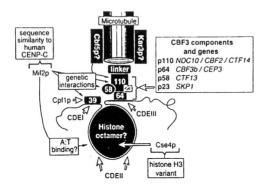

PLUTA et al. 1995 - Hypothetical model of the budding yeast kinetochore. Three conserved DNA elements (CDEI, CDEII and CDEIII), a polypeptide (CENP), histone proteins and other components are shown.

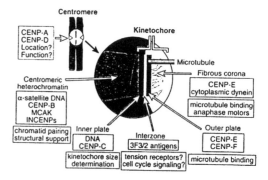

PLUTA et al. 1995 - Components of the mammalian centromere. Different types of binding proteins (CENP) and the alfa-satellite DNA are included.

THE CENTROMERE - A PANDORA'S BOX OF UNEARTHED PROPERTIES -- 2

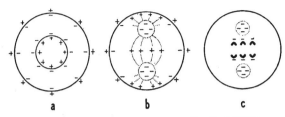

LILLIE 1905, 1911 - Diagrams representing the interpretation of chromosome movements based on electrostatic forces.

LIMA-DE-FARIA 1955 -

Chromosomes of *Allium cepa* showing centromeres divided while the two sister chromatids are still held together by the arms. Upper figures, camera lucida drawings; lower figures photomicrographs of the same chromosomes.

GOYANES and SCHVARTZMAN 1981 -

Electron micrograph of a human diplochromosome confirming the division of the centromere prior to that of the arms.

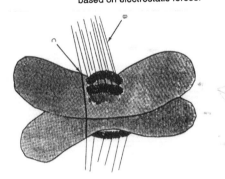

BRINKLEY and STUBBLEFIELD 1970 -
Diagrammatic representation of microtubules attached to kinetochore filaments.

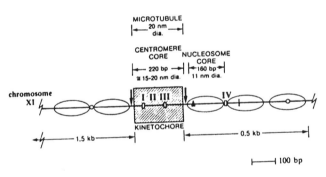

CARBON 1984 - Organization of the yeast centromere in terms of base pairs (bp) and nanometers (nm) with the sequenced DNA segments I, II and III.

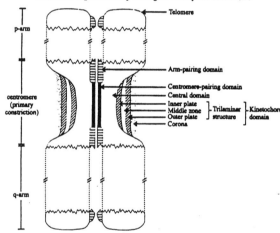

CHOO 1997 - Diagram of the structural domains of the human centromere-kinetochore complex. These include a centromere-pairing domain distinct from the sister chromatid arm pairing domains and a central region divided into several zones.

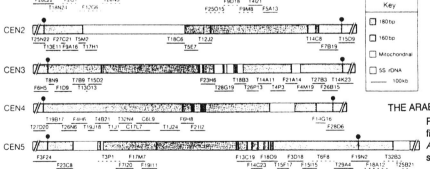

THE ARABIDOPSIS GENOME INITIATIVE 2000 -

Predicted centromere composition of the five chromosomes (CEN1 to CEN5) of *Arabidopsis*, which have been partially sequenced.

The telomere
Not just a terminus station

From the beginning of cytology the ends of chromosomes were found to behave in particular ways. As early as 1921, J. Gelei described the movement of chromosome ends during meiosis to a polar region of the nucleus. Later this observation was confirmed by several authors who pictured the polarization of chromosome ends at pachytene (C.D. Darlington 1936) and their attachment to the nuclear envelope (E. Gilson et al. 1993).

Origin of the telomere concept – The main pioneering contribution of H.J. Muller was the experimental production of mutations in *Drosophila*. Under the action of X-rays, not only did the genes mutate, but the chromosomes were broken into fragments. Surprisingly not all chromosome pieces survived. Only those that, besides a centromere, had retained their natural ends could do so. This finding led H.J. Muller to write in 1940: "The end genes thus constitute permanent chromosome structures, distinguished from all other genes by their constantly monopolar character; they have been denoted as 'telogenes' (or, for the visible bodies, 'telomeres') both by Darlington and myself and by Haldane. Similarly persistent structures are the 'centromeres' (as Darlington calls them)".

It turned out that what applied to *Drosophila*, was valid for the chromosomes of other animals and of plants. The telomere concept became established on a firm basis.

Structure – The structure of telomeres remained in part elusive. From the beginning, H.J. Muller thought that the telomere consisted of a single terminal band that sealed the very end of a *Drosophila* polytene chromosome. But soon other cytologists found that these chromosomes could lack several terminal bands without having difficulties to survive (M. Demerec and M.E. Hoover 1936). This was the first indication that the telomere was a region much longer than originally envisaged. The analysis of telomeres in plant chromosomes at pachytene (in which the chromomere pattern is most distinct) led to the finding that a chromosome end could be divided into as many as eight segments. This compound pattern was found in six species belonging to the *Solanaceae*, *Labiatae*, *Liliaceae*, and *Gramineae* (A. Lima-de-Faria 1958). This multiple nature of the telomere has been confirmed by deletions obtained on the terminal region of chromosome 9 of maize (B. McClintock 1944) and on the terminal region of chromosome 10 also of maize (T. Mello-Sampayo 1966). One of the accessory chromosomes of rye shows great variation of the telomere region in populations originating from different regions of Europe and Asia (A. Lima-de-Faria 1963).

Location – The location of telomeres would seem to be obvious, but like that of centromeres, it turned out to be far from regular.

One of the first cases was reported by B. McClintock (1941). She found a permanent healing of median segments in maize chromosomes, an indication that other regions than the natural ends manifested monopolarity. Since then interstitial telomeres have been found in many species, one of them the Chinese hamster whose genome contains 18 cytologically detectable interstitial telomeric sequences (P. Slijepcevic et al. 1997). The interstitial localization of telomeric DNA has also been described in the Indian and Chinese muntjacs (H. Scherthan 1990, C. Lee et al. 1993) as well as in humans (R.A. Wells et al. 1990).

Molecular organization – The molecular organization of telomeres has now been studied extensively. The sequencing of the natural ends is known for organisms as simple as protozoans, such as *Euplotes* and *Paramecium*, as well as for tomato and humans (D. Kipling 1995). Telomeres are often composed of repeated short, simple sequences, such as $(TTAGGG)_n$. The conservation is impressive, this sequence being the same in *Trypanosoma brucei* and *Homo sapiens*. The telomere DNA is associated with various types of telomere-binding proteins. These are different in somatic and germ-line cells. TRF1, TRF2 and KU are found in the soma, but human sperm telomeres contain a specific binding-protein complex that is a variant of histone H2B (A.A. Gineitis et al. 2000).

The replication problem – Circular chromosomes, such as those found in bacteria, can replicate without natural ends. Linear chromosomes face an obstacle during replication, every round of synthesis will result in a small loss of a terminal sequence due to the specific properties of DNA polymerase. This difficulty has been circumvented by several mechanisms, one of them being the special enzymatic activity of telomerase, an enzyme that adds a telomeric sequence onto (in this case) the upper strand of a DNA molecule. The other strand is replicated by DNA polymerase. The combined effect is the formation of a telomeric sequence (D. Kipling 1995).

The single-stranded portion of the telomere folds back on itself forming special G–G base pairs (H. Lodish et al. 1995).

Movement on the spindle – Most surprising is the finding that in several species, such as rye and maize, telomeres can function as centromeres, being active on the spindle. The introduction of an abnormal chromosome 10 in maize, which carried a large block of heterochromatin, resulted in the telomeres of several chromosomes assuming a mobile activity even stronger than that of centromeres (M. Rhoades 1952). Recently telomeres were found to function as centromeres in yeast cells, leading the chromosomes during their movements to the poles (Y. Chikashige et al. 1994).

In many animal species, and in some plant species, the centromere has a strictly terminal position assuming the function of a telomere. It is remarkable that they can substitute each other both structurally and functionally.

The emergence of new telomeres – Another similarity between telomeres and centromeres is that telomeres can also arise anew. In the chromosomes of nematodes, such as *Ascaris*, chromosome fragmentation is a normal process, which results in the formation of minute new chromosomes with their own telomeres (M.J.D. White 1954). Chromosome fragmentation and telomere formation also occur in ciliated protozoa. Programmed chromosome rearrangements lead to the formation of polytene chromosomes, destruction of non-genic DNA, *de novo* telomere addition to gene-sized molecules leading to the formation of the mature macronucleus (D. Kipling 1995). Interstitial telomere sequences have been described in the assembly of deer chromosomes (H. Scherthan 1990 and C. Lee et al. 1993).

Unanswered questions – Much remains to be elucidated concerning: 1) the origin of new telomeres, 2) the causes of the similarities between centromeres and telomeres, 3) the internal limits of telomeres, 4) the way telomeres affect the function of genes located either in their vicinity or far away from them (D.De Bruin et al. 2000), 5) the relationship that seems to exist between telomere length and cell aging (D. Shore 1997).

THE TELOMERE - NOT JUST A TERMINUS STATION -- 1

GELEI 1921 - Polarization of chromosome ends during prophase of meiosis in *Dendrocaelum*.

DARLINGTON 1936 - Polarization of chromosome ends in *Chorthippus*.

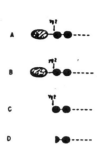

McCLINTOCK 1944 - Observed deficiencies in the chromomeres at the end of the short arm of chromosome 9 of maize.

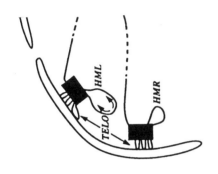

GILSON et al. 1993 - Telomeres and heterochromatic domains of *S. cerevisiae* associated at the nuclear periphery.

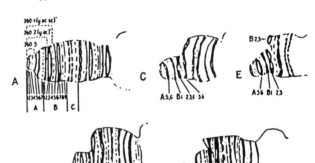

DEMEREC and HOOVER 1936 - Deficiencies at the tip of the X chromosome of *Drosophila melanogaster* involving several bands and known genes.

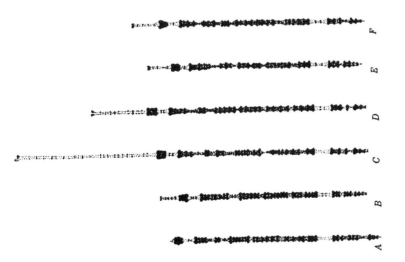

LIMA-DE-FARIA 1963 - Variation in chromomere pattern at the terminal region of the B chromosome of rye from 6 populations of Europe and Asia (A-F).

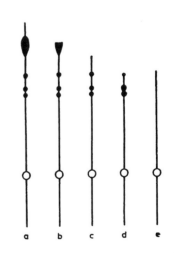

MELLO-SAMPAYO 1966 - Variation in chromomere pattern at the end of the long arm of chromosome 10 of maize.

THE TELOMERE - NOT JUST A TERMINUS STATION -- 2

RHOADES 1952 - Camera lucida drawing of anaphase II of meiosis in maize showing the active movement of the chromosome ends to the spindle poles. This happens when a chromosome 10 with a large knob is present in the cells.

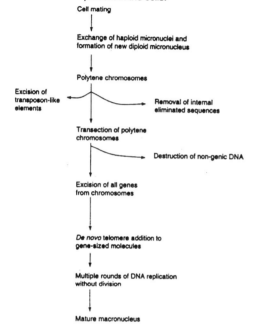

KIPLING 1995 - Programmed chromosome rearrangements during macronuclear development in hypotrichous ciliates in which new telomeres are added.

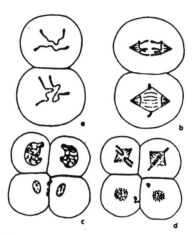

WHITE 1954 - The regular fragmentation of the large chromosomes in the cells of *Parascaris* leading to the formation of minute chromosomes with own telomeres.

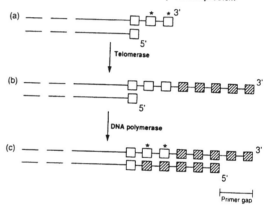

KIPLING 1995 - Telomere replication by telomerase.

Sequence	Species
TTTTGGGG	*Euplotes, Oxytricha, Stylonychia*
TTGGGG	*Glaucoma, Tetrahymena*
TTGGGG and TTTGGG	*Paramecium*
TTTAGGG and TTCAGGG	*Plasmodium*
TTAGGG	*Trypanosoma brucei*
TAGGG	*Giardia lamblia*
$(TG)_{1-6}TG_{2-3}$	*Saccharomyces cerevisiae*
$TTACAG_{1-8}$	*Schizosaccharomyces pombe*
TTAGGGGG	*Cryptococcus neoformans*
TTAGGG	*Podospora anserina*
TTAGGG	*Fusarium oxysporum*
TTAGGG	*Neurospora crassa*
ACGGATGTCTAACTTCTTGGTGT	*Candida albicans*
TTAGGG	*Didymium iridis*
TTAGGG	*Physarum polycephalum*
AG_{1-8}	*Dictyostelium discoideum*
TTAGGC	*Ascaris lumbricoides*
TTGCA	*Parascaris univalens*
TTAGG	*Bombyx mori*
HeT-A, TART (retroposons)	*Drosophila melanogaster*
TTAGGG	Humans, mice
TTTAGGG	*Arabidopsis thaliana*
TT(T/A)AGGG	Tomato
TTTTAGGG	*Chlamydomonas reinhardtii*

KIPLING 1995 - Terminal repeat DNA sequences found in organisms as diverse as protozoa and humans.

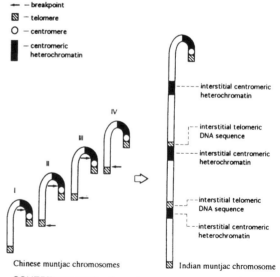

SCHERTHAN 1990, LEE et al. 1993 - Assembly of an Indian muntjac chromosome from four Chinese muntjac chromosomes resulting in the occurrence of interstitial telomeric DNA sequences.

The nucleolus organizer
Nothing in the cell is comparable to it

The nucleolus – An exotic chromosome formation
What exactly is the nucleolus as a cell organelle? Nothing in the cell is comparable to it. It is not present in the cytoplasm, where most organelles are located, but it is found within the nucleus. Moreover, it is not a structure floating freely in the nuclear sap, but surprisingly it is attached to the chromosome body.

At prophase of mitosis and meiosis chromosomes have a well delimited body surface. At certain stages they build loops but they do not display side outgrowths or other kinds of accessory bodies. The exception is the formation of the nucleolus.

Like one of the gods of Greek mythology, the chromosome carries on its back a heavy and large sphere, as did the giant Atlas who was compelled to support the World on his shoulders. Still more remarkable is that this sphere has no outer membrane to keep its shape. The nucleus and the cell have their own well organized membranes, that maintain their form, but the nucleolus does not possess an outer limiting structure. How it maintains this spherical shape is unknown. Only future work on its molecular content will reveal the source of this unique situation.

The term nucleolus springs from the initial observation that within the nucleus there was another minor nucleus, hence the diminutive designation of nucleolus. It soon turned out that this was not a smaller nucleus but an organelle with its own properties connected directly with the chromosomes (J. Gelei 1921, S.L. Frolowa 1935). This relationship was elucidated by E. Heitz (1931) who showed that the number of "secondary constrictions" observed in metaphase chromosomes corresponded to the number of nucleoli present in interphase nuclei. In one of her first classic contributions B. McClintock (1934) produced a translocation between chromosomes 6 and 9 of maize, which showed that the nucleolus organizer region was a multiple structure that could be divided into two partially functional segments. The association of the nucleolus with protein synthesis was suggested by T. Caspersson (1950) and by 1957, W.S. Vincent could isolate clean nucleoli in large amounts, a prerequisite for biochemical analysis. Soon the general properties of ribosomes were established and these were found to be the site of protein synthesis (J.C.H. De Man and N.J.A. Noorduyn 1969). The nucleolus then became recognized as the site of ribosome preformation (R.P. Perry 1969) and by this time O.L. Miller and B.R. Beatty (1969) isolated the "Christmas Trees" consisting of DNA from the nucleolus to which were attached nascent ribosomal RNA molecules in progressive stages of synthesis. What became subsequently evident was that the genes for 5S RNA (which is also an integral part of ribosomes) were located not at the nucleolus organizer region but in other chromosomes. This shows that distantly located genes (even in different chromosomes) can collaborate in the production of a well defined molecular edifice. There are at present more examples of this situation and the question that arises is what kind of molecular guidance is available in these molecules that directs them in a precise way to a specific cell site and that obliges them to enter into a specific molecular assembly.

The processing of pre-ribosomal RNA in the nucleolus is now better understood as well as the successive steps of association of the 18S, 28S and 5S ribosomal RNAs with proteins until their final transfer into the cytoplasm in the form of ribosomal subunits (H.F. Noller 1991, C.S. Pikaard 2000).

Origin and structure of the nucleolus organizer
The nucleolus organizer has in metaphase chromosomes a structure that is indistinguishable from that of the centromere. Both regions appear, at this mitotic stage, as constrictions of the chromosome body consisting of weakly stained fibrillae. But at pachytene, the stage at which the chromomere pattern is most distinct, the centromere continues to consist of fine fibrillae containing small chromomeres, whereas the nucleolus organizer usually consists of large chromomeres (forming a "knob") to which the nucleolus is attached.

The origin of the nucleolus is unknown, it does not exist in bacteria, it is a eukaryotic invention. Its origin may have been connected with the sequestration of the chromosomes in the nucleus. This obliged new solutions connected with the formation of proteins and their transfer between the cytoplasm and the nucleus.

Molecular organization
As the techniques from bacterial genetics became available to cytologists working with eukaryotic chromosomes, a search was made to localize the genes for ribosomal RNA. In science one needs to follow the simplest path and ribosomal RNA could be easily extracted from eukaryotic cells. F.M. Ritossa and S. Spiegelman (1965) were the first to isolate this RNA and to localize its genes in *Drosophila* chromosomes. They found that their site

was the nucleolus organizer. Soon this finding was confirmed for other animals and plants and the genes for 18S and 28S, which were accompanied by large "spacer" regions, were sequenced and their promoters identified (R.M. Learned et al. 1987).

Location of the nucleolus organizer
Information on the location of the actual site of nucleolus formation, was collected from 506 species ranging from algae to humans. This study disclosed a distribution pattern with special features. In 86.6% of the cases investigated the nucleolus organizer was located on the short arm, i.e. the ribosomal genes tended to occur in the arm where the centromere-telomere distance was shortest. Within the short arms the genes did not occur at random but established themselves near the telomere and along a line which was parallel to it. This resulted from the fact that, as the centromere-telomere distance increased, the ribosomal genes were successively displaced in such a way that a location close to the telomere was maintained. The other arm of the nucleolar chromosome, which did not contain a nucleolus, also had non-random features. As the nucleolar arm increased in length the arm without the nucleolus also tended to become longer. This was an indication that the second telomere was also involved in some form. Algae and Bryophyta did not differ in this respect from higher mammals and humans. The distribution of the ribosomal genes was so well defined within the centromere-telomere field that it could be predicted by means of a straight-line equation (A. Lima-de-Faria 1973, 1979).

Experiments were carried out to test the maintenance of gene territory in deer species in which the chromosome number varies from 6 to 70 chromosomes (diploid). The analysis of their DNA, with restriction enzymes, and in situ hybridization, revealed that the genes, were not mixed during these drastic chromosomal transformations, but maintained their territory in relation to the position of telomeres and centromeres (A. Lima-de-Faria et al. 1986, L. Frönicke and H. Scherthan 1997).

Unanswered questions
Many molecular mechanisms remain to be elucidated concerning the nucleolus. One of them is its disintegration at pro-metaphase of mitosis and meiosis and its reassembly at telophase. Do the nucleolar molecules, delivered directly into the cytoplasm, have a role in the subsequent chromosome movements? Recently it has been discovered in yeast, that the proteins concentrated in the nucleolus also intervene in the pachytene "check point", *i.e.* in the mechanism that prevents chromosome missegregation (G.S. Roeder and J.M. Bailis 2000).

What molecular interaction obliges the nucleolar chromosome to get rid of this heavy burden as it prepares to move to the daughter cells? The secondary constriction replaces the nucleolus (at its site of formation) during metaphase of mitosis. What is the function of this region of the chromosome that, at this stage, takes the form of an extremely thin chromosomal thread?

The location of the nucleolus close to telomeres raises many questions concerning the interactions between specialized chromosome regions.

THE NUCLEOLUS ORGANIZER -
NOTHING IN THE CELL IS COMPARABLE TO IT-- 1

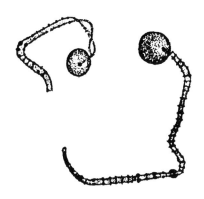

GELEI 1921 - Two chromosomes with a terminally located nucleolus at meiosis in *Dendrocaelum*.

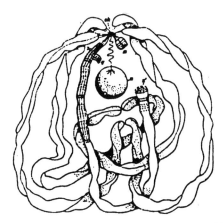

FROLOWA 1935 - The chromosome complement of *Drosophila funebris* in a salivary gland nucleus, showing the nucleolus suspended from the chromocenter.

HEITZ 1931 - The relationship between the number of secondary constrictions present in metaphase chromosomes and the number of nucleoli seen in the interphase nucleus of six plant species.

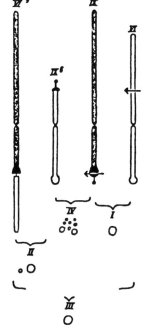

McCLINTOCK 1934 - Translocation between chromosomes IX and VI of *Zea mays* resulting in the division of the nucleolus organizer into two partially functional regions. The nucleoli are shown below the chromosome combinations (I, II, III, IV).

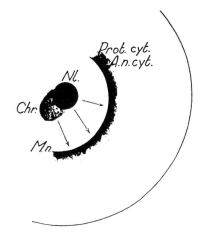

CASPERSSON 1950 - Mechanism of protein synthesis according to which proteins migrated from the nucleolar heterochromatin to the nuclear membrane (Nl, nucleolus).

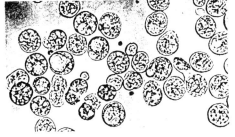

VINCENT 1957 - Nucleoli isolated from starfish oocytes.

THE NUCLEOLUS ORGANIZER -
NOTHING IN THE CELL IS COMPARABLE TO IT -- 2

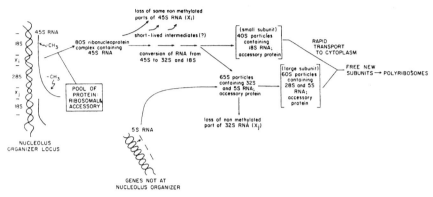

PERRY 1969 - Schematic representation of the biosynthesis of ribosomes in a typical eukaryotic cell.

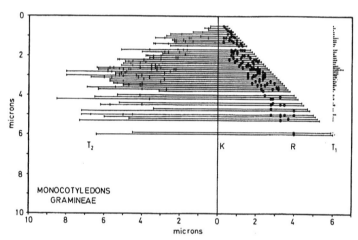

LIMA-DE-FARIA 1973 - Location of the nucleolus organizer (black circles) in 97 species of the family Gramineae (K, centromere, T1 and T2 telomeres).

NOLLER 1991 - Binding states of transfer RNA during translation of messenger RNA on a ribosome consisting of two subunits 30S and 50S.

	70 S ribosomes		80 S ribosomes	
1. M.W.	2.7×10^6		4×10^6	
2. RNA content	65%		45%	
3. Protein content	35%		55%	
4. Basic proteins with M.W.	25,000		12,000–25,000	
5. Diameter	140–270 Å		220–300 Å	
6. Optimum Mg^{2+} concentration for maximum amino acid incorporation	10–15 mM		1.5 mM	
7. Inhibition of protein synthesis by antibiotics	Inhibition		No inhibition	
8. Subunits	30 S	50 S	40 S	60 S
M.W.	0.9×10^6	1.8×10^6	1.3×10^6	2.6×10^6
M.W. RNA	0.6×10^6	1.2×10^6	0.6×10^6	1.7×10^6
S value RNA	16	23	18	28

MAN and NOORDUYN 1969 - General properties of ribosomes.

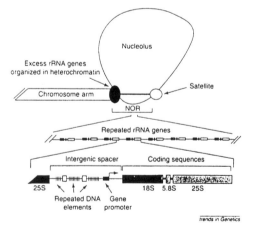

PIKAARD 2000 - Organization of a typical nucleolus organizer region.

PART V

NO CHROMOSOME CAN FUNCTION OUTSIDE A CELL

Cytoskeleton
A disgusting artifact became an important cell edifice

We tend to forget that no chromosome can function outside a cell. Even a virus chromosome, although it may exist outside this unit, in the form of a virion particle, cannot replicate or function outside the cell. For this reason a short comment on the cell's main organelles seems necessary to better understand the chromosome's dependence on the cell's organization.

Living cells when studied under the microscope, as early as 1838, revealed cytoplasmic movements in the form of streams which flowed between a network of filaments (M.J. Schleiden 1838). Such a reticular formation was seen time and again in fixed material (O. Bütschli 1878) and the living cell continued to display the same pattern (M. Heidenhain 1897). But researchers, working with chromosomes, tried to obtain preparations in which the cytoplasm was as transparent and as structureless as possible, otherwise they could not count and analyse the chromosomes properly. This was a scientific period in which every deviating structure was labeled as an "artifact". The result was that, as late as 1950, the "reticular formation" of the cytoplasm was described as protein precipitation due to bad fixation and as such a structure to be disregarded (J.R. Baker 1950). It was only 30 years later that the biochemical and ultrastructural analysis of the cell, started to disclose that one was dealing with a cell framework, which was called the cytoskeleton, and which turned out to consist of an intertwining of microfilaments containing actin monomeres and other proteins (H. Curtis 1983, E. Lazarides 1994, R.D. Goldman 1995).

The cytoskeleton is now known, among other things, to regulate chloroplast motility in plant cells (M.K. Kandasamy and R.B. Meagher 1999). It has also been shown that it participates in animal cell motility, morphogenesis and cell secretion (B. Safiejko-Mroczka and P.B. Bell 2001, M.L. Munoz et al. 2001).

CYTOSKELETON.
A DISGUSTING ARTIFACT BECAME AN IMPORTANT CELL EDIFICE.

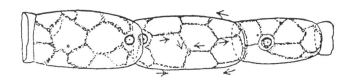

SCHLEIDEN 1838 - First description of cytoplasmic streaming.

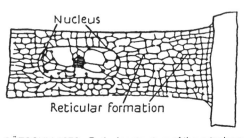

BÜTSCHLI 1878 - Reticular structure of the cytoplasm in a fixed cell.

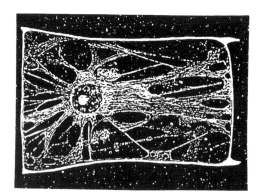

HEIDENHAIN 1897 - Network of filaments in a living cell.

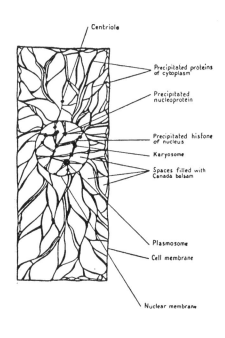

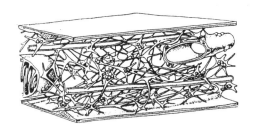

CURTIS 1983 - Microtrabeculae form the bulk of the cytoskeleton together with microfilaments.

BAKER 1950 - Cell with "artifacts" due to precipitation of proteins.

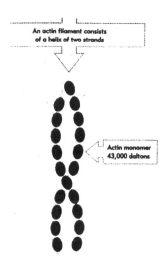

LAZARIDES 1994 - Composition of actin microfilaments.

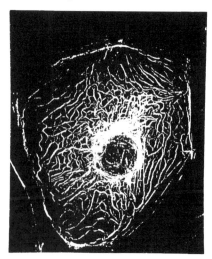

GOLDMAN 1995 - A cell showing its keratin filament network.

Nuclear envelope
The nucleus disclosed its outer structure

From the onset the shape of the nucleus was found to be usually spherical. For a long time this was accepted as a natural condition which did not demand further investigation. Later it was recognized that the nucleus was encircled by a membrane, but this seemed to have no obvious structure. Moreover, an intriguing process occurred at every cell division: this membrane was broken down at pro-metaphase, releasing the chromosomes into the cytoplasm, and then suddenly it was rebuilt at telophase imprisoning the chromosomes again in the nuclear sap.

The introduction of the electron microscope was most valuable in this respect. The nuclear envelope turned out to contain many pores that allowed molecular transit, in and out of the nucleus, and these pores showed a complex morphology (H. Moor 1967, B.J. Stevens and J. André 1969). The exact dimensions of the annuli were determined in different species and their organization became better known (C.W. Akey and M. Radermacher 1993). What happened at telophase during the reassembly of the nuclear envelope could also be reconstructed in detail (G.P. Vigers and M.J. Lohka 1991).

The light microscope as well as the E.M., had revealed, a long time ago, that chromosomes were many times attached to the nuclear envelope (L.R. Cleveland 1938, M.J. Moses and J.R. Coleman 1964). The functional significance of this position is still not fully understood. Another area that has been poorly investigated is the molecular composition of what is called the nuclear sap (H. Busch et al. 1964). A great deal remains to be learned in this area; the chemical messages that occur within the nucleus, between chromosomes, await investigation.

NUCLEAR ENVELOPE.
THE NUCLEUS DISCLOSED ITS OUTER STRUCTURE.

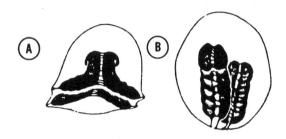

CLEVELAND 1938 - The chromosomes of the flagellate *Spirotrichonympha* are attached to a persistent nuclear envelope (A and B).

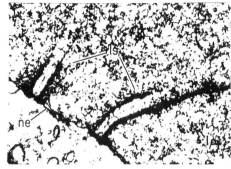

MOSES and COLEMAN 1964 - Termination at the nuclear envelope (ne) of two chromosomes in a rat spermatocyte (displaying synaptonemal complexes).

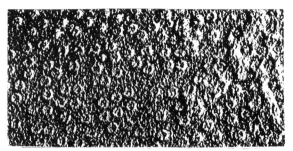

MOOR 1967 - Nuclear envelope of a salivary gland cell from *Drosophila* (frozen-etched preparation).

Amino acid	Percentage of total moles
Alanine	9.4
Arginine	5.5
Aspartic acid	10.1
Cystine	0.7
Glutamic acid	10.0
Glycine	8.4
Histidine	3.8
Isoleucine	5.9
Leucine	9.5
Lysine	8.7
Methionine	1.3
Phenylalanine	4.1
Proline	5.5
Serine	6.3
Threonine	5.2
Tyrosine	2.1
Valine	7.4

BUSCH et al. 1964 - Amino acid composition of proteins of the nuclear sap.

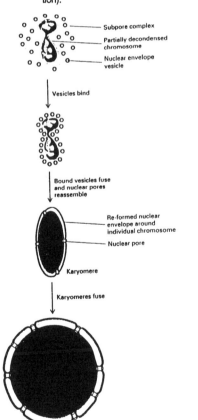

VIGERS and LOHKA 1991 - Assembly of the nuclear envelope during telophase.

STEVENS and ANDRE 1969 - Diagram of a nuclear pore as viewed in a section normal to the nuclear surface.

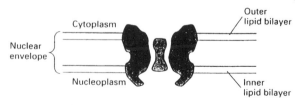

AKEY and RADERMACHER 1993 - Structure of the nuclear pore complex.

Centriole
An enigmatic cell invention

This organelle was described in the early days of the century as being associated with spindle formation and with chromosome separation during cell division (M. Jörgensen 1913, E.B. Wilson 1925). The replication of this minute organ and its migration to the regions called the spindle poles, became evident during the 1960s. Its internal structure, consisting of a highly ordered and symmetric array of microtubules, was well established at the same time (A. Ross 1968, E.J. DuPraw 1970). In subsequent years the electron microscope picture has only received minor modifications, but it has been complemented by molecular knowledge of the various categories of proteins involved in the assembly of microtubules (H. Lodish et al. 1995).

The centriole remains an enigmatic organ for the simple reason that it has for a long time displayed properties, that defied our simplistic concept of the cell. Centrioles are known to possess remarkable capacities: they replicate and they generate other organized structures in the cytoplasm. Firstly, replication is generally considered to be performed only by DNA and RNA molecules, but there is no definitive evidence that these molecules are part of centrioles. It actually turns out that centrioles replicate during G_1 *i.e.* just before the S phase, whereas chromosomes duplicate their DNA during the S phase. Moreover the replication of centrioles occurs, not in the nucleus, but in the cytoplasm. Secondly, in higher plants the spindle fibers converge on a cytoplasmic pole where no centrioles are visible, yet chromosome movements occur as well as in animal cells in which centrioles are present. Thirdly, the centriole can give rise to cilia. This means that centrioles lead to the assembly of other structures in a precise form and at a precise time. In other words our knowledge of the self-assembly mechanisms which govern cell organization and function remains highly obscure.

Microtubule proteins have been extensively studied as well as the nuclear genes responsible for their production. These proteins are an integral part of cilia and flagella and the mitotic apparatus is a microtubule machinery (S.J. Doxsey et al. 1994). The structural and functional interrelations between these cell components await clarification.

CENTRIOLE.
AN ENIGMATIC CELL INVENTION.

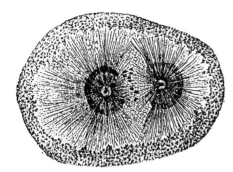

JÖRGENSEN 1913 - Oocyte of *Piscida* showing asters irradiating from centrioles.

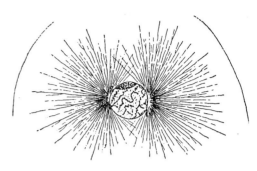

WILSON 1925 - Asters in the sea-urchin *Toxopneustes*.

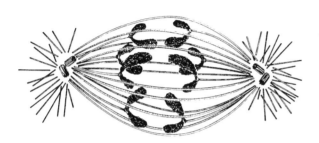

DUPRAW 1970 - Mitotic apparatus at anaphase. A centriole pair lies at each pole.

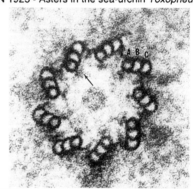

ROSS 1968 - Cross-section of a human centriole. The 9X3 tubules seen in the E.M. at high magnification.

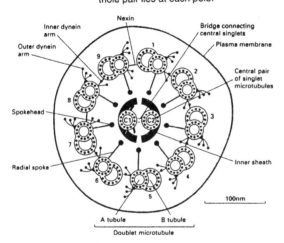

LODISH et al. 1995 - Cross-section view of an idealized flagellum showing all major structures.

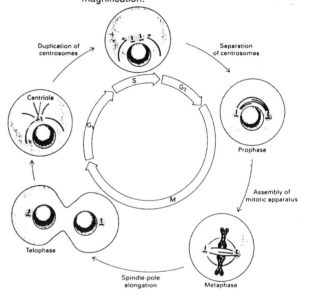

LODISH et al. 1995 - The centriole cycle. Centrioles duplicate during the G1 of the cell cycle.

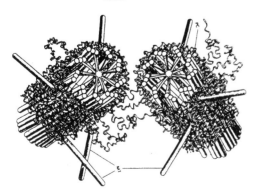

← DOXSEY et al. 1994 - Microtubule-organizing center with a pair of centrioles. These are connected by fibers and are surrounded by a pericentriolar (PC) matrix which contains gamma-tubulin. The ends of microtubules (MT) are embedded in the matrix.

Endoplasmic reticulum and Golgi apparatus
The building of membranes permitted molecular reactions to occur in defined sequences

During the early decades of genetics, cytologists tended to think of the chromosomes as main inhabitants of the cell and to forget about the cytoplasm. This attitude had its origin in the fact that the cytoplasm had to be, as transparent and structureless as possible, if one was going to identify, and delimit, the chromosomes properly. But chromosomes cannot function without a cytoplasm: (1) At metaphase and anaphase they get rid of the nucleus and function in this cell compartment. (2) Protein synthesis is a process in which both chromosomes and cytoplasm participate. (3) No chromosomes can function outside the cell, even the chromosomes of viruses need to return to the cell cytoplasm and nucleus to replicate.

At the turn of the century C. Garnier (1897) had called attention to the "ergastoplasm" suggesting its role in cell secretion. The development of the electron microscope, and the ultracentrifugation techniques, redirected the attention of cytologists from the nucleus to the cytoplasm. The work of K. Porter, G.E. Palade and F.S. Sjöstrand in the 1960s led to a breakthrough in this area. The term endoplasmic reticulum was coined for the system of cisternae present in the cytoplasm of cells. The reticulum may be continuous with the plasma membrane or the nuclear envelope. When the outer surfaces of its membranes are coated with ribosomes it is called rough-surfaced, otherwise it is said to be smooth-surfaced (P.J. Goldblatt 1969). The endoplasmic reticulum is part of the whole system of membranes found in the cell and the main structural component of protein synthesis.

The use of amino acids labelled with the radio-isotope tritium permitted following accurately the transit of molecules in the cytoplasm and the interaction between the endoplasmic reticulum and the Golgi apparatus (J.D. Jamieson and G.E. Palade 1967). This cell organelle, described initially by the Italian scientist C. Golgi, is also composed by a series of membranes that form cisternae, but these have a different arrangement and lack ribosomes (P. Favard 1969). The Golgi collects the molecular products originating in the endoplasmic reticulum and imprints an address on many proteins allowing them to reach an exact destination in the cell (A.P. Pugsley 1989). Both organelles are main agents in cell secretion (P. Favard 1969).

The building of membranes allowed the cell, in its early phase, to carry out molecular reactions in defined sequences. As these structures were formed molecules could attach to a permanent substrate in a precise sequence. Membranes were primary instruments in bringing order to the chemical events occurring in the cytoplasm. Protein synthesis, with its many steps, is just one example of such processes.

Striking is the finding that membranes do not occur within the nucleus. They do not exist in the nuclear sap; and the nucleolus and the chromosome, are devoid of them. Although the cytoplasm and the nuclear contents are mixed at every metaphase and anaphase, this structural distinction is maintained. We have at present no idea of the causes of this situation.

ENDOPLASMIC RETICULUM AND GOLGI APPARATUS.
THE BUILDING OF MEMBRANES PERMITTED MOLECULAR REACTIONS TO OCCUR IN DEFINED SEQUENCES.

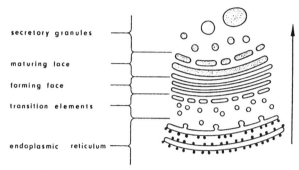

FAVARD 1969 - Diagram illustrating the relation between the endoplasmic reticulum and a dictyosome leading to the formation of secretory granules.

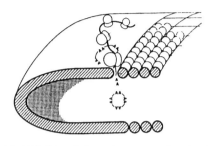

GOLDBLATT 1969 - Schematic cross section of end of cisterna of rough endoplasmic reticulum showing a growing chain of amino acids (AA).

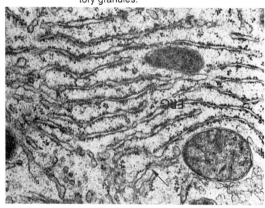

GOLDBLATT 1969 - Stacks of rough-surface cisternae (EGR) which are part of the endoplasmic reticulum.

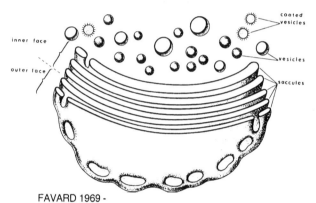

FAVARD 1969 -

Diagram illustrating the ultrastructure of a dictyosome which is the main component of the Golgi apparatus. The dictyosome consists of a stack of flattened saccules. Numerous smooth vesicles and occasional coated vesicles are found in the vicinity of the inner face of the stack.

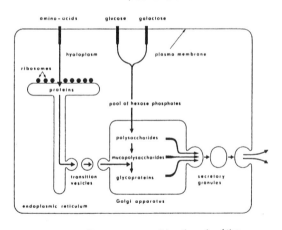

FAVARD 1969 - Diagram summarizing the role of the Golgi apparatus in the cell.

JAMIESON and PALADE 1967 - Localization of radioactivity after injection of ^3H-leucine in a pancreatic cell from guinea pig. Successive movement of the labeling from the endoplasmic reticulum to the Golgi apparatus (dictyosomes) and finally to the zymogen granules.

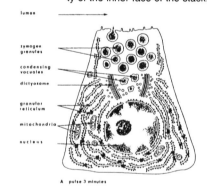

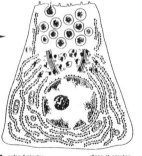

Cell membrane and cell wall
The cell became an individualized entity

The chromosome does not exist in a vacuum, but all its building blocks, like those of any other cell organelle, have their origin in the outer environment. It is the cell wall, in plants, and the cell membrane, in both the plant and animal kingdoms, which perform the main task of controlling the flow of outer atoms and molecules into the cell's machinery as well as their transit to the exterior (i.e. endocytosis and exocytosis).

X-ray analysis of plant cell walls revealed, as early as 1929, the molecular configuration of the molecules of glucose that build large micellae and the position of their connecting oxygen atoms (W. Seifriz 1929, O.L. Sponsler 1929). Also before the advent of the electron microscope, A. Frey-Wyssling (1939), described the sub-microscopic structure of the cellulose fibers that build the wall of plant cells.

The parallel orientation of microfibrils of cellulose could be clearly seen in the electron microscope (E. Frei and R.D. Preston 1961) and the actual growth of the cell wall became the object of a study by P.A. Roelofsen (1965). P. Albersheim (1980) presented a model of the cell wall in which hemicellulose molecules were hydrogen bonded to the surface of cellulose microfibrils which in turn were linked to microfibrils of pectin.

The cell membrane was initially compared to a colloidal emulsion (G.H.A. Clowes 1916). By 1935 it was evident that it consisted of two protein layers, one facing the exterior and the other the interior of the cell, both being separated by a lipid part (J.F. Danielli and H.A. Davson 1935). In the 1950s and 1960s more knowledge was added which furnished molecular details on the architecture of these two main components (J.F. Danielli 1954). The transport of substances across the membrane, as well as its chemical composition, became well established (W. Stockem and K.E. Wohlfarth-Bottermann 1969 and G. Guidotti 1972). A more complex relationship between the phospholipid bilayer and the peripheral proteins, as well as the association with other molecular components, has become evident in later years (H. Lodish et al. 1995).

The molecular components of cell membranes can self-assemble, just as other cellular components self-assemble into ribosomes and nucleosomes, but much remains to be elucidated, at the atomic level, of the mechanisms involved in these processes (H. Lodish and J.E. Rothman 1979).

CELL MEMBRANE AND CELL WALL.
THE CELL BECAME AN INDIVIDUALIZED ENTITY --1

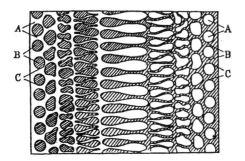

CLOWES 1916 - Colloidal emulsion resembling a cell membrane.

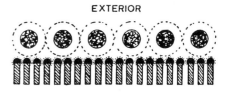

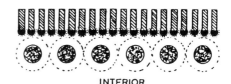

DANIELLI and DAVSON 1935 - One of the first molecular models of a membrane.

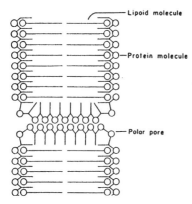

DANIELLI 1954 - Diagram of the cell membrane with polar pores.

Membrane	Percentage by Weight		
	Protein	Lipid	Carbohydrate
Myelin	18	79	3
Plasma membrane:			
Human erythrocyte	49	43	8
Mouse liver	44	52	4
Ameba	54	42	4
Chloroplast spinach lamellae	70	30	0
Halobacterium purple membrane	75	25	0
Mitochondrial inner membrane	76	24	0

GUIDOTTI 1972 - Chemical composition of some purified membranes.

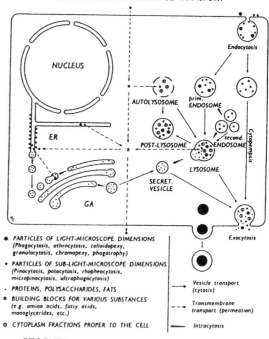

STOCKEM and WOHLFARTH-BOTTERMANN 1969 - Membrane and substance transport involving endocytosis and exocytosis.

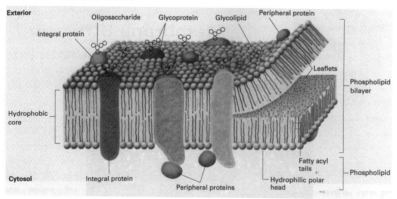

LODISH et al. 1995 - Schematic diagram of a typical biological membrane.

CELL MEMBRANE AND CELL WALL.
THE CELL BECAME AN INDIVIDUALIZED ENTITY --2

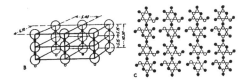

SPONSLER 1929 - Cell wall showing chains consisting of glucose residues (black circles, oxygen atoms; white circles, carbon atoms).

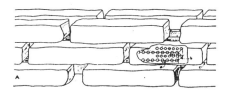

SEIFRIX 1929 - Cell wall structure, as indicated by X-ray analysis, consisting of large micellae.

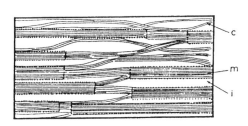

FREY-WYSSLING 1939 - Submicroscopic structure of a cellulose fiber (c, cellulose chains; m, crystalline cellulose; i, intermicellar space).

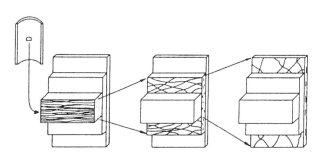

ROELOFSEN 1965 - Successive stages of the growth of a young wall.

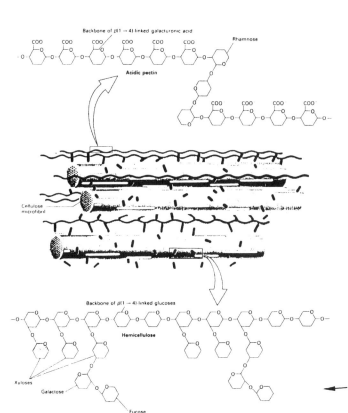

FREI and PRESTON 1961 - Electron micrograph of cell wall of a green alga showing microfibrils of cellulose.

ALBERSHEIM 1980 - Structure and interconnections of the major polymers in the primary cell wall.

PART VI

SPECIFIC TYPES OF CHROMOSOMES

Chromosomes of viruses
An early or a late form of chromosome?

When the chromosomes of plants and animals were discovered, viruses were not known, and it took a long time before their existence was even suspected. They started to be recognized as agents of disease that were so small that they passed through the filters that retained bacteria. As late as 1939 they were not even included in dictionaries of biology.

One had to wait until the 1960s to find them mentioned in treatises on genetics as "infectious, subcellular and ultramicroscopic particles representing potentially pathogenic agents". At this time, their role as agents of many diseases was far from being well established.

The breakthrough came in 1935 when W.M. Stanley obtained crystals of tobacco mosaic virus. Soon H. Ruska and collaborators (1939) could display the first pictures of virus particles in the electron microscope. The year 1952 saw the publication of the classic work by A.D. Hershey and M. Chase in which they demonstrated that it was the DNA of the virus that was the carrier of genetic information. This DNA led to the production of new virus particles in the infected bacteria. Since then it has been known that viruses consist of protein and contain either RNA or DNA as genetic material. They can be crystallized into extremely regularly shaped particles but they cannot reproduce autonomously. They need bacterial or eukaryotic cells, to obtain a progeny, by diverting the cell's molecular machinery to their own use. It is this process that led to the extensive study of the viral chromosomes during the 1950s and 1960s.

The electron micrographs of tobacco mosaic virus could show clearly the filaments of its RNA projecting out of the coat protein (R.C. Hart 1955). This led to a model in which the helical arrangement of the RNA is surrounded by protein monomers (A. Klug and D.L.D. Caspar 1960). The viruses which infected bacterial cells, the bacteriophages, were also studied in the electron microscope following an osmotic shock. The highly packed chromosome of T2, consisting of DNA, exploded out of the virus head (A.K. Kleinschmidt et al. 1962). A diagram of the organization of a phage particle was produced by E.A. Evans (1956). The base composition and nucleic acid content of mammalian viruses were summarized by J.H. Subak-Sharpe (1969).

The operons of bacteriophage lambda were well established by 1970 (J.D. Watson 1970) and a map of its chromosome showing the physical position of some genes became available later (R.W. Old and S.B. Primrose 1980). One of the most striking results was obtained in T4 phage in which the whole process of the assembly of the virus was elucidated. As a result the genes responsible for each assembly step (represented by numbers) were identified (W.B. Wood and R.S. Edgar 1967; W.B. Wood 1973). Better electron micrographs have been obtained in later years of tobacco mosaic virus particles (R. Williams 1988) and the electron density map of a protein disk has been published by A. Klug (1988).

Viruses change their genetic make-up rapidly. This poses at present a serious problem from the pathologic point of view. Another area of future research is the elucidation of their relation to the eukaryotic chromosome. It is not known how these nucleic-acid – protein assemblies acquired a genetic independence but remained prisoners of the cell. Moreover, only when we know better, how eukaryotic chromosomes are organized as functional units, will we be in a position to get an insight into the mechanisms that direct virus stability and variability.

A pressing problem is the understanding of the origin of viruses. Did they accompany the emergence of the cell or were they formed later? Their evolutionary histories and relationships to contemporary organisms remain unknown. This is due to the following. 1) Viral genomes are remarkably diverse. 2) The sequences of viral DNAs have diverged past the point of recognition. 3) Bacteriophages as a group are not so similar to one another. In a study of the evolutionary origin of viruses R.W. Hendrix (1999) has placed them next to their host on the tree of life, to emphasize the difficulties involved in the unraveling of their origin.

CHROMOSOMES OF VIRUSES.
AN EARLY OR A LATE FORM OF CHROMOSOME? -- 1

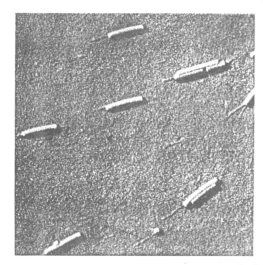

HART 1955 - Electron micrograph of tobacco mosaic virus; filaments of RNA projecting out of the virus.

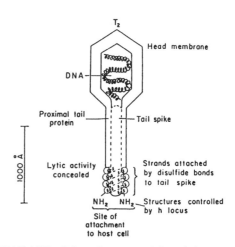

EVANS 1956 - Schematic representation of phage particle.

Virus	(G+C)%	Molecular weight of genome $\times 10^6$
Single-stranded RNA		
EMC	47	3
Polio	46	2
Single-stranded DNA		
MVM	41	1.7
H1	45	1.7
Kilham RV	43	1.7
Double-stranded DNA		
Polyoma	48	3
SV$_{40}$	41	3
Shope papilloma	47	5
Human papilloma	41	5
Adenovirus type 2	57	23
Vaccinia	35	160
Equine rhinopneumonitis	55	70
Herpes	68	70
Pseudorabies	74	70
Host cells		
Human spleen	41	1.8×10^{12}

SUBAK-SHARPE 1969 - Comparison of the base composition and nucleic acid content of mammalian viruses.

KLUG and CASPAR 1960 - The helical arrangement of RNA and coat protein in a tobacco mosaic virus.

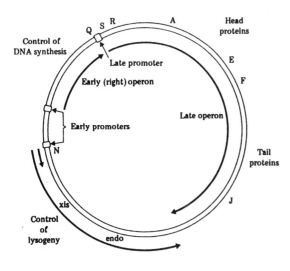

WATSON 1970 - Operons of bacteriophage lambda.

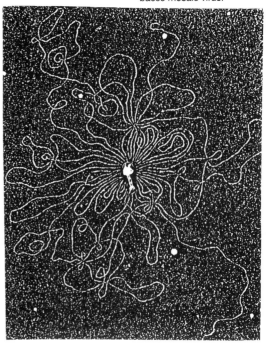

KLEINSCHMIDT et al. 1962 - Electron micrograph of a T2 bacteriophage after osmotic shock, showing the DNA outside the head.

Chromosomes of bacteria
Nearly naked DNA could become independent

The emergence of bacteria is nearly synonymous with the origin of life on earth. For this reason bacterial chromosomes are considered if not the most primitive, quite close to that state. But is it really so? Bacteria have had 3.5 billion years at their disposal to evolve and during this time they have been able to populate the most adverse environments, living at temperatures above 100 °C and surviving on oil deposits. The chromosomes present in today's living bacteria may differ appreciably from those found in the first prokaryotes. Several mechanisms exist in their chromosomes that indicate an advanced condition, such as the presence of two types of chromosomes. The main circular large chromosome contains most genes. Minor additional chromosomes, called plasmids or episomes, have a few additional genes and replicate autonomously. Both types consist of DNA.

Microbial genetics has a long history to which are associated the names of many pre-eminent colleagues. It is their research that laid the foundations of molecular genetics, which later extended to the molecular cytogenetics of higher organisms, and which finally led to present-day molecular biology of the cell. B. Ephrussi, S.E. Luria, M. Delbruck, G.W. Beadle, J. Lederberg, E.L. Tatum, F. Jacob, J. Monod and S. Spiegelman are among those who made fundamental contributions.

From the beginning *Drosophila* was the organism of choice for genetic experiments, and as a consequence, became the first organism to be used in the analysis of biochemical reactions involving genes. In 1935 G.W. Beadle and B. Ephrussi studied the genetics of eye pigments in this fly and in 1940 G.W. Beadle collaborating with E.L. Tatum extended these biochemical studies to the fungus *Neurospora*. This work resulted in the formulation of the one gene – one enzyme hypothesis. In the meantime S.E. Luria and M. Delbruck recognized that spontaneous mutations occurred in bacteria. The year was 1943 and their paper is considered to represent the birth of bacterial genetics. This work was soon followed by J. Lederberg's and E.L. Tatum's classic experiment (1946), in which they demonstrated the occurrence of genetic recombination in biochemical mutants from *E. coli* strains.

From the beginning it was not known if bacteria had chromosomes, for they had no nucleus. It turned out that they had chromosomes harboured directly in their cytoplasm and that these were circular, having no centromeres and no telomeres. These chromosomes consisted of DNA. This made it possible to label them with tritiated molecules and to study their replication by a combination of autoradiography with electron microscopy (J. Cairns 1963). The whole DNA content of *Hemophilus influenzae* was spread on an electron microscope film following the treatment of the bacterium by an osmotic shock, its total length was found to be 832 microns (L.A. MacHattie et al. 1965).

The genes of *E. coli* were soon located along its circular chromosome and by 1964 about 500 genes had been mapped (A.L. Taylor and M.S. Thoman 1964). The natural position of the chromosome filaments inside the bacterium was shown in *Bacillus subtilis* by W. van Iterson (1969). The similarities and differences between the chromosomes of bacteria and of higher organisms were summarized by M. Westergaard (1964) but some of this evidence has since then been superseded.

The bacteria were among the first organisms to create accessory chromosomes in addition to their basic genetic complement. In plants and animals these take the form of extra chromosomes that are carried at the side of the normal complement. Their presumed ancestors are the episomes of bacteria which consist of small DNA rings harbouring a few genes (L.E. Hood et al. 1975).

E. coli became one of the ideal objects of genetic studies. F. Jacob's and J. Monod's classic work on the *lac* operon (1961) led to its DNA sequencing (R.C. Dickson et al. 1975). The transition states in the folding and unfolding of the *E. coli* chromosome have been depicted by A. Worcel and E. Burgi (1972) and its process of segregation has been represented by G.S. Stent (1971) and by R.T. Wheeler and L. Shapiro (1997).

In the last few years some bacteria have been found to have linear instead of circular chromosomes (C.W. Chen 1996). The question arises whether this represents the dawn of the linear eukaryotic chromosome. Another feature, which may also represent a transition state between pro- and eukaryotes, is the description in bacteria of a mitotic-like machinery of active chromosome segregation (M.E. Sharpe and J. Errington 1999).

CHROMOSOMES OF BACTERIA.
NEARLY NAKED DNA COULD BECOME INDEPENDENT --1

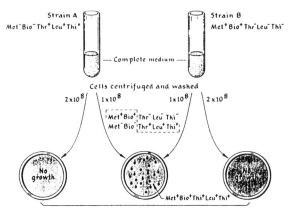

LEDERBERG and TATUM 1946 - Schematic summary of Lederberg and Tatum's classic genetic cross of two auxotrophic mutant strains of E. coli K12.

Bacteria	Higher organisms
A single, circular DNA chromosome. No centromere or telomeres.	Several nucleoprotein chromosomes. Telomeres probably always present; an individualized centromere usually present in each chromosome.
No mitotic apparatus, division by DNA replication. No nuclear membrane.	Mitotic apparatus present, mitosis follows on DNA replication after an interval of time. Nuclear membrane present.
Partial transfer of genetic information (merozygosis) by sexual reproduction, transformation and transduction.	Complete transfer of genetic information (holozygosis) by sexual reproduction. Transformation and transduction not found.
Reduction (haploidization) through crossing-over in the merozygote whereby one of the two products of recombination, being lethal, is eliminated.	Reduction through meiosis, all four products of which are initially (at least) viable.

WESTERGAARD 1964 - Comparison between cytogenetic mechanisms in bacteria and higher organisms.

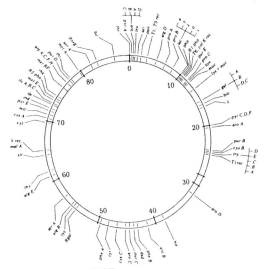

TAYLOR and THOMAN 1964 -
A simplified genetic map of E. coli. The inner cycle indicates time, in minutes, of genetic transfer.

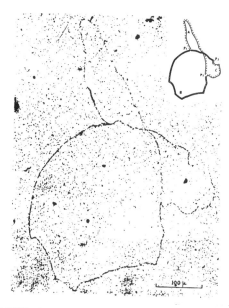

CAIRNS 1963 - Autoradiograph of the intact chromosome of E. coli labeled with H^3-thymidine. The cartoon shows the portions of double- and single-stranded labeling.

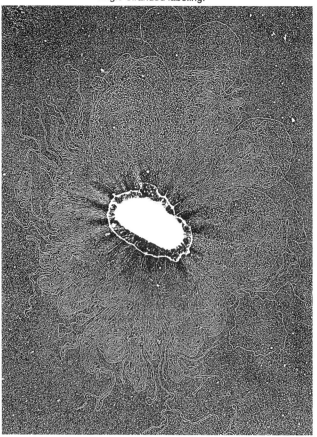

MacHATTIE et al.1965 - Electron micrograph of the bacterium H. influenzae after osmotic shock, revealing a DNA chromosome about 832 microns long.

CHROMOSOMES OF BACTERIA.
NEARLY NAKED DNA COULD BECOME INDEPENDENT --2

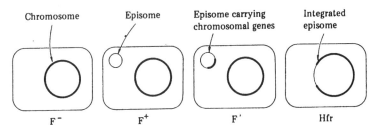

HOOD et al. 1975 - Genetic elements and mating types of *E. coli*.

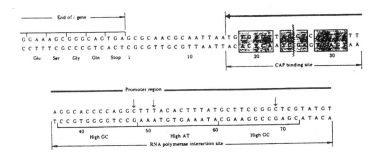

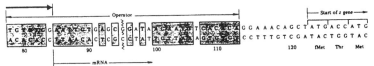

DICKSON et al. 1975 - Nucleotide sequence of *lac* operon control elements.

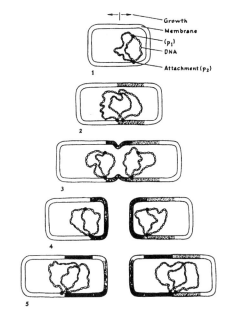

STENT 1971 - Mechanism of distribution of the bacterial chromosome.

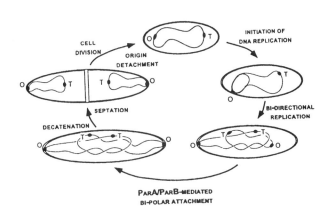

WHEELER and SHAPIRO 1997 - A model of bacterial chromosome segregation.

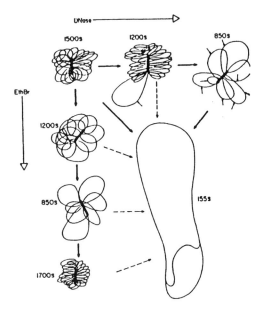

WORCEL and BURGI 1972 - Coiling and uncoiling of the *E. coli* chromosome, from a compact folded supercoil to an unfolded chromosome.

Chromosomes of mitochondria
Intruders invaded the cell

Mitochondria were initially considered by chromosome researchers to be disgusting little dots that appeared in the cytoplasm of plant and animal cells. They were only a source of confusion, since they obscured the delimitation of chromosomes in dividing cells. As late as the 1950s the main objective of chromosome fixation and staining, was to obtain a totally clear cytoplasm where no minor organelles interfered with the identification of the nuclear chromosomes. But for biologists, interested in other cell functions, the small dots, seen mainly outside the spindle apparatus at metaphase of mitosis, became the focus of attention (W. Robyns 1924). During the 1930s and 1940s researchers, such as A. Guilliermond and G. Mangenot (1941) studied the mitochondria attentively. These organelles, of about 1 micron in diameter, were found to be most fragile. They were easily destroyed by acetic acid and alcohol fixatives. It was ascertained that they had the same shape and chemical constitution (mainly protein and lipid) in plant and animal cells, but their origin, as well as their structure and function, remained elusive. Besides, it was not clear what relationship they had with chloroplasts or with other cell organelles.

After World War II the knowledge of mitochondria changed radically. The cell fractionation methods of A. Claude and collaborators and the electron microscopy of G.E. Palade, K.R. Porter and F.S. Sjöstrand elucidated both the structure and function of these organelles. By the 1960s it was clear that the primary function of mitochondria was providing the cell with ATP (adenosine triphosphate) and that they had other secondary functions. However, as late as 1961 A.B. Novikoff wrote that "mitochondria lack DNA and possess little, if any RNA". By 1967 this doubt was dispelled, the DNA had been isolated and visualized within this organelle (I.B. Dawid and D.R. Wolstenholme 1967, P. Borst and A.M. Kroon 1969, M. Rabinowitz et al. 1969). The DNA formed circles that were isolated in EM preparations and the DNA of the kinetoplast, a mitochondrion found in trypanosomes, built a great number of rosettes (D.L. Fouts et al. 1975). By treating the cell of *Euglena gracilis* with specific dyes the large nuclear mass of DNA and that of mitochondrial DNA could be visualized (Y. Huyashi and K. Veda 1989).

The genetics of human and of yeast mitochondria progressed rapidly during the 1980s due to the efforts of several groups among them: P. Borst and L.A. Grivell 1981, A. Chomyn et al. 1986 and A. Tzagoloff and A. Myers 1986. The mitochondrial genes and their products involved ribosomal RNAs, ribosomal proteins, cytochrome c oxidase, ATPase and others (V.K. Eckenrode and C.S. Levings 1986, C.S. Levings and G.G. Brown 1989). The sites of transcription and translation of some mitochondrial proteins were depicted by C.W. Birky, Jr. (1976).

It came as a surprise when the genetic code, which had been found to be universal, suddenly turned out to differ in mitochondria, the codons having other properties than in the nuclear DNA (S. Anderson et al. 1981, P. Borst 1981, P.S. Covello and M.W. Gray 1989 and others). Exceptions to the code also occurred in several ciliates (protozoa).

Lately the structure and function of mitochondria have become well defined (H. Lodish et al. 1995) but much remains to be studied, as their molecular pathology, connected with several human diseases, starts to be investigated (F.S. Sjöstrand 1999).

CHROMOSOMES OF MITOCHONDRIA.
INTRUDERS INVADED THE CELL -- 1

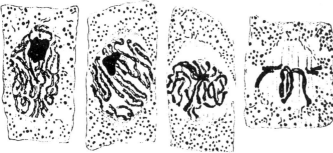

ROBYNS 1924 - Mitochondria located in the cytoplasm in fixed material.

GUILLIERMOND and MANGENOT 1941 - Mitochondria seen in the living cell's cytoplasm.

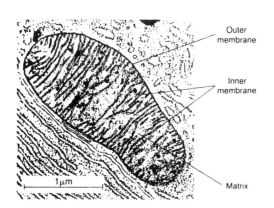

PORTER (Published 1995 but from the 1960s) - The mitochondrion reveals its internal structure in the E.M.

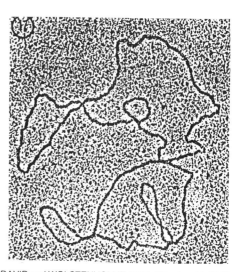

DAVID and WOLSTENHOLME 1967 - Mitochondria DNA from *Xenopus*.

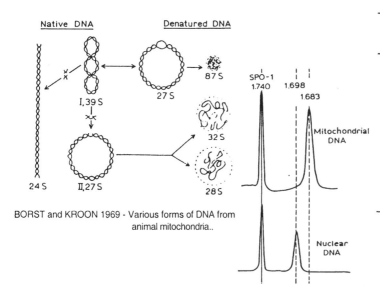

BORST and KROON 1969 - Various forms of DNA from animal mitochondria.

RABINOWITZ et al. 1969 - Optical density of nuclear and mitochondrial DNA.

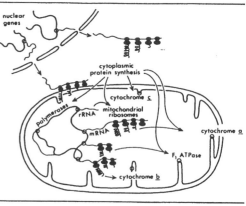

BIRKY JR. 1976 - Sites of transcription and translation of some mitochondrial proteins.

CHROMOSOMES OF MITOCHONDRIA.
INTRUDERS INVADED THE CELL -- 2

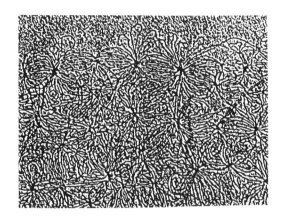

FOUTS et al. 1975 - E.M. picture of kinetoplast DNA.

BORST and GRIVELL 1981 - CHOMYN et al. 1986 - TZAGOLOFF and MYERS 1986 - The genetic organization of human (inner rings) and yeast (outer rings) mitochondrial DNA.

	Mammal	Yeast	Fungus (Neurospora)	Plant
Size (thousands of base pairs)	14–18	78	19–108	200–2500
rRNAs				
Large subunit	16S	21S	21S	26S
Small subunit	12S	15S	15S	18S
5S RNA	–	–	–	+
No. of tRNAs	22	23–25	23–25	~30
Ribosomal protein (var-1)	–	+	+	?
Subunits 1, 2, 3 of cytochrome c oxidase	+	+	+	+
Apocytochrome b subunit of CoQH$_2$–cytochrome c reductase	+	+	+	+
F$_0$ ATPase				
Subunit 6	+	+	+	+
Subunit 8	+	+	+	+
Subunit 9	–	+	–	+
Subunit α of F$_1$ ATPase	–	–	–	+
No. of NADH-CoQ reductase subunits	7	0	6	6

ECKENRODE and LEVINGS 1986 - Mitochondrial genes
LEVINGS and BROWN 1989 - and their products.

Codon	Standard Code: Nuclear-Encoded Proteins	Mitochondria				
		Mammals	Drosophila	Neurospora	Yeasts	Plants
UGA	Stop	Trp	Trp	Trp	Trp	Stop
AGA, AGG	Arg	Stop	Ser	Arg	Arg	Arg
AUA	Ile	Met	Met	Ile	Met	Ile
AUU	Ile	Met	Met	Met	Met	Ile
CUU, CUC, CUA, CUG	Leu	Leu	Leu	Leu	Thr	Leu

ANDERSON et al. 1981 -
BORST 1981 -
COVELLO and GRAY 1989 -
Alterations in the standard genetic code in mitochondria.

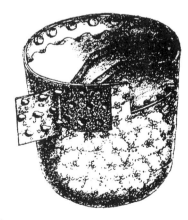

LODISH et al. 1995 - Three-dimensional diagram of a mitochondrion cut longitudinally.

SJÖSTRAND 1999 - Drawing based on electron micrographs of structure of mitochondria following partly denatured proteins.

Chromosomes of chloroplasts
Additional genomes entered the cell

During the 1930s the studies of plant cells revealed minute structures in their cytoplasm. These were distinct organelles considered to be responsible for photosynthesis. Their origin remained uncertain, and still more equivocal, was their relationship with other types of plastids and mitochondria. There was, however, no doubt that chloroplasts were of the utmost importance in cell function (A. Guilliermond 1912).

The staining of a cell component by the Feulgen test method was cytochemical demonstration that it contained DNA. This had, time and again, been shown to apply for nuclear chromosomes. However, no one believed that chloroplasts could contain DNA or genetic information. When Y. Chiba in 1951, using the Feulgen staining technique, demonstrated that DNA occurred in chloroplasts, his work was dismissed as an artifact. Chloroplasts were considered to be too small and too indistinct when observed under the light microscope. His result could not be taken seriously.

One had to wait for the improvement of biochemical methods and of electron microscope techniques to demonstrate the presence of DNA in chloroplasts and to be able to unravel their complex membrane system. In 1955 E. Steinman and F. Sjöstrand got clear pictures of the grana and its membrane structure in the EM. A series of researchers published photographs of the chloroplast structure (D. Branton and R.B. Park 1967, J.A. Lauretis and E. Kitajima 1969, L.K. Shumway 1970).

The analytical ultracentrifuge had become a standard instrument in the isolation of DNAs. This allowed separating the main peak of the nuclear DNA from the smaller peak of chloroplast DNA (R. Sager and M. Ishida 1963). This DNA could be spread on grids and seen directly in the electron microscope (C.L.F. Woodcock and H. Fernandez-Moran 1968). By this time chloroplast ribosomes from bean leaves could be visualized in the EM (V.I. Bruskov and M.S. Odintsova 1968).

The genes encoded by chloroplast DNA from the liverwort *Marchantia polymorpha* became a good example of the many genetic functions present in this organelle, most of them involved in the production of the proteins essential for photosynthesis (K. Ohyama et al. 1986). But the photosynthetic complexes of the thylakoid membranes could be coded by the chloroplast or nuclear genomes or by both. The chloroplast DNA was found to be a circle, like the chromosome of bacteria, and it contained inverted repeats and short and long single copy regions. The double-stranded DNA molecule consists of 120 to 160 kilobases, harbouring circa 130 genes (M. Sugiura 1992, B. Lewin 1994).

The 3-dimensional organization of a chloroplast has been well established: its "thylakoid" membranes are fused into "grana" which reside in a matrix, the "stroma". All the chlorophyll is contained in the thylakoid vesicles (H. Lodish et al. 1995).

The chloroplast genome had similarities to both prokaryotic and eukaryotic DNAs. In 1972 G.H. Pigott and N.G. Carr found that the ribosomal RNAs from cyanobacteria hybridized with DNA from the chloroplasts of *Euglena gracilis*. This genetic homology supported an earlier assumption that chloroplasts were the descendants of endosymbiotic bacteria. Moreover, the protein synthetic apparatus of chloroplasts displayed resemblances to the bacterial apparatus, many of the ribosomal proteins being homologous to those of *E. coli*.

The resemblance with eukaryotic DNA was as follows: The chloroplast genome coded for circa 50 proteins including RNA polymerase and some ribosomal proteins. The organelle genes were transcribed and translated by the organelle machinery. Some chloroplast genes carried introns, a feature common to nuclear genes. The physicochemical properties of higher plant chloroplast, blue-green algal and bacterial DNA were summarized by S.D. Kung (1977).

Much remains to be learned about chloroplast genetics, as well as about the biogenesis of chloroplasts, since many of its key steps are not understood at the molecular level. Chloroplasts, like mitochondria, grow by expansion and then fission. This means that we are highly ignorant of the mechanism that preserves their genetic order during organelle multiplication. Moreover, the plastid genome has an extremely high ploidy level. Over 100 chloroplasts are found in a single tobacco leaf cell and each plastid may contain about 100 copies of the genome. This means that there are 10,000 plastid genomes per cell (A.J. Bendich 1987, S. Ruf et al. 2001).

CHROMOSOMES OF CHLOROPLASTS.
ADDITIONAL GENOMES ENTERED THE CELL --1

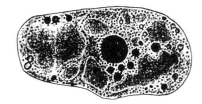

GUILLIERMOND 1912 - Study on the origin of mitochondria and chloroplasts.

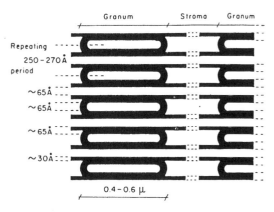

STEINMAN and SJÖSTRAND 1955 - Schematic representation of the components of a chloroplast (granum and stroma) based on E.M. studies.

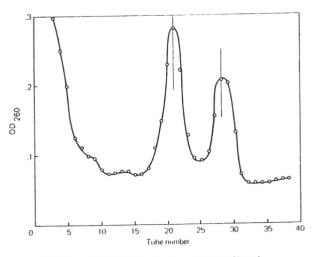

SAGER and ISHIDA 1963 - Sedimentation profiles of DNA from the alga *Chlamydomonas*. Whole cell DNA (major peak), chloroplast DNA (minor peak).

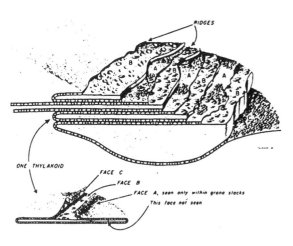

BRANTON and PARK 1967 - Interpretation of the thylakoids of a chloroplast.

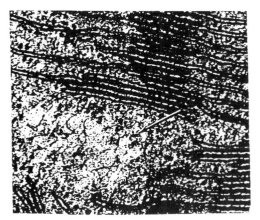

SHUMWAY 1970 - Electron micrograph of maize chloroplast showing photosynthetic membranes and DNA region (arrow).

Physicochemical properties of higher plant chloroplast, blue-green algal and bacterial DNA

Properties	Chloroplast	Blue-green algal DNA	Bacterial DNA
Buoyant density (g cm^{-3})	1.697 ± 0.004	1.705 ± 0.010	1.716 ± 0.025
GC content	38 ± 2	48 ± 9	50 ± 25
Renaturation	Extensive	Extensive	Extensive
Renaturation rate	Rapid	Slow	Slow
Repeated sequence	Absent	Absent	Absent
5-Methylcytosine	Absent	Present	Present
DNA content (×10^{-14} g)	0.2–0.9	3.0–3.6	1.9–2.8
Dry weight (%)	0.04–0.15	0.40–1.70	0.80–4.20
Conformation	Circular	Circular	Circular
Histones	Absent	Absent	Absent
DNA organization	25 Å fibrils	25 Å fibrils	25 Å fibrils
Satellite band	Absent	Absent	Absent

KUNG 1977 - Comparison of the properties of chloroplast DNA with that of bacteria and blue-green algae.

CHROMOSOMES OF CHLOROPLASTS.
ADDITIONAL GENOMES ENTERED THE CELL --2

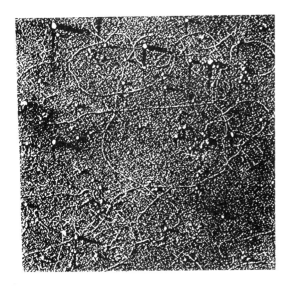

WOODCOCK and FERNANDEZ-MORAN 1968 - Dispersed chloroplast DNA seen in the E.M.

Genes	Description of Gene Product
rRNA (duplicated in inverted repeats IR_A and IR_B)	16S, 23S, 4.5S, and 5S RNAs
tRNA	37 tRNAs
RNA polymerase	Homologous to *E. coli* RNA polymerase:
rpo A	subunit α
rpo B	subunit β
rpo C_1	subunit β'
rpo C_2	subunit β'
Ribosomal proteins	
rpl	50S subunit: 8 proteins
rps	30S subunit: 11 proteins
Proteins essential for photosynthesis	
rbcL	Large subunit of ribulose 1,5-bisphosphate carboxylase
psaA, psaB	Chlorophyll *a*-binding proteins in photosystem I
psbA	Photosystem II D1 protein
psbB, psbC	Chlorophyll *a*-binding proteins in photosystem II
psbD	Photosystem II D2 protein
psbE, psbF	Cytochrome b_{559}
psbG	Photosystem II G protein
atpA, atpB, atpE	Subunits α, β, and ϵ of F_1 ATPase
atpF, atpH, atpI	Subunits 1, 3, and 4 of F_0 ATPase
petA	Cytochrome *f*
petB	Cytochrome b_6
petD	Subunit 4 of cytochrome *b/f* complex
Genes predicted by amino acid sequence homology	
ndh1, ndh2, ndh3, ndh4, ndh4L, ndh5, ndh6	Homologous to subunits of the human mitochondrial NADH-CoQ reductase complex
frxA, frxB, frxC	Homologous to ferredoxin
Others	More than 28 unidentified open reading frames

OHYAMA et al. 1986 - Genes encoded by chloroplast DNA from a liverwort, *Marchantia polymorpha*.

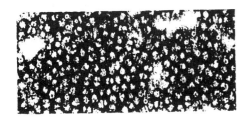

BRUSKOV and ODINTSOVA 1968 - Ribosomes of chloroplasts from bean leaves.

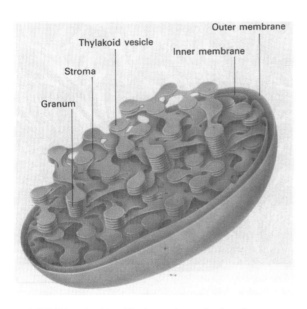

LODISH et al. 1995 - The internal organization of a chloroplast.

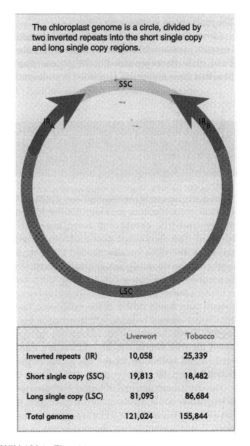

	Liverwort	Tobacco
Inverted repeats (IR)	10,058	25,339
Short single copy (SSC)	19,813	18,482
Long single copy (LSC)	81,095	86,684
Total genome	121,024	155,844

LEWIN 1994 - The chloroplast genome is a circle containing inverted repeats plus short and long single copy regions.

PART VII

THE ANTITHETICAL PROPERTIES OF THE CHROMOSOME

Physico-chemical processes are antithetical

Antithetical phenomena, or structures, are defined as having opposing or contrasting properties (N. Webster 1976).

Physicists have long been confronted with the duality exhibited by most physical phenomena in the form of negative forces opposing positive ones, of attraction counterbalancing repulsion, of expansion inhibiting contraction and other phenomena with an antithetical nature.

As experimental physics came of age after the beginning of the 1900s a plethora of elementary particles was discovered. One main finding was that they were characterized by antithetical properties. The proton was positive counterbalancing the negative electrons. The neutrino could be left-handed but also right-handed. Even the same elementary particle could occur in two opposing states. The discovery of the electron (with its negative charge) later led to the prediction of the occurrence of another particle with the same mass but having a positive charge. This was soon identified and was named positron. Later the antiproton was also discovered. This evidence was aptly summarized by H. Alfvén (1966) in his book "Worlds-Antiworlds, Antimatter in Cosmology" and recently by H.R. Quinn and M.S. Witherell (1998, see Part IX, page 181).

Cells and chromosomes consist only of the elementary particles that are part of all matter. It is thus not surprising that the antithetical nature of the physico-chemical processes was transferred to the cell and the chromosome.

The cell also consists of molecules which are left-handed and right-handed. The amino acids, the DNAs and the sugars, occur both in the left-handed and the right-handed forms. One of the forms of each type of macromolecule tends to predominate in the cell, just as one of the forms of each elementary particle predominates in the construction of matter.

Moreover, in the living system molecules counteract each other or have opposing effects. Enzymes activate and inhibit the function of other molecules. Antigens are counteracted by antibodies. Insulin stimulates anabolic processes and inhibits catabolic reactions. The same protein can activate and repress transcription, as is the case with the yeast protein RAP1 (A. H. Brand et al. 1987). A long list could be added of antithetical molecular events which are a permanent part of the cell's function.

The chromosome has not escaped from this conflicting situation. Its genetic processes and functions contain many antithetical situations. The terms anti-codon, antiparallel RNA and antisense RNA have become familiar. Specific genes function as activators whereas others counterbalance their effect as suppressors. Oncogenes lead to malignant growth whereas anti-oncogenes regulate normal tissue development.

The chromosome's enormous rigidity is accompanied by an equally impressive plasticity. These antithetical properties take various forms and their mapping may allow us to better understand and control chromosome evolution.

THE CHROMOSOME'S RIGIDITY

Maintenance of organization
The protozoan versus the human chromosome

The emergence of the eukaryotic chromosome

The big jump was between the chromosome of bacteria and that of eukaryotes. Once the chromosome left the cytoplasm to become sequestered in a nucleus, basic differences arose, both in organization and behaviour, with unexpected consequences.

The emergence of the eukaryotic chromosome resulted in a series of new events. (1) From circular the chromosome became mainly linear. (2) Histones combined with its DNA, a marriage that has lasted over one billion years. (3) The nuclear envelope encircled the chromosome during interphase and other mitotic stages, but allowed the chromosomes to get out of this prison during metaphase and anaphase bathing in the cytoplasm as they did in prokaryotes. (4) Telomeres were invented, a feature that put exact limits to the newly arisen linear structure. (5) Centromeres were also created, allowing a more efficient separation of the chromosomes to daughter cells. (6) The regular reduction in chromosome number was also added, resulting in a means of producing gametes with half of the organism's main chromosome number. (7) Large chromosome rearrangements became the order of the day allowing new DNA combinations. (8) Crossing over also made its inroads during meiosis as well as into mitosis. (9) Chromosome pairing, accompanied by specific structures called synaptonemal complexes, was also introduced as an accessory system of order. (10) Last, but perhaps not least the nucleolus emerged, a most unique appendage hanging from the chromosome like a flag from its pole. (11) Introns became a general feature of structural genes. (12) RNA splicing was improved as an extra way of changing the DNA message without altering the DNA itself. (13) Structural changes, instead of occurring only longitudinally along the chromosome, were allowed to occur vertically, piling copies over copies of gene sequences, building what has been called polyteny and amplification.

Many of the above-mentioned properties were recently discovered, and hence the eukaryotic chromosome may have in store many other surprises concerning its own capacities of innovation. Did the chromosome produce all these solutions alone, or was the cell in which it is immersed a partner in its transformations? No clear answer can be given to this question at present, but a partial one is available. There are two separate lines of cell evolution: one occurring when the chromosome has been locked in a single cell (as in a protozoa), the other when the chromosome became part of a metazoan organism in which it occupied billions of cells within a single individual. The question that arises is: Did the chromosome of higher organisms acquire quite different properties or did it remain like the protozoan chromosome?

One of the features of the eukaryotic chromosome that is seldom pointed out, is the impressive constancy of organization that it has displayed since the origin of eukaryotic organisms. The earliest protozoa appeared about 1.5 billion years ago and were followed by multicellular animals and plants about 1.0 billion years ago (J.W. Schopf 1987, K.H. Nealson and P.G. Conrad 1999).

If one compares the chromosome of a unicellular organism with that of a mammal, such as a human being, one is confronted with an essentially identical structure. Both have the same: (1) main features, (2) chemical composition, (3) structural organization and (4) functional processes.

Comparison between human and protozoan chromosomes

1) The chromosomes of both groups are essentially similar. When studied, in the light or electron microscope, a human chromosome at mitosis or meiosis, cannot be easily distinguished from that of a protozoa. This is especially clear when examining the 26 chromosomes of *Barbulanympha ufalula* at diakinesis drawn by L.R. Cleveland (1954).

2) Both chromosome types consist of DNA associated with histones. The chromatin of the macronucleus of a *Holosticha* sp. that has been spread by the cytochrome c method shows, in the electron microscope, gene-sized DNA molecules building rosettes and free DNA molecules, which are like those of any metazoa (K. Murti and D.M. Prescott 1986). The histones of the protozoan *Tetrahymena* are similar to those of yeast, peas and cows, as shown by the molecular analysis carried out by M.A. Gorovsky (1986) in which the percentage differences between these proteins are tabulated for H4, H3, H2A and H2B.

3) Centromeres. These specialized regions responsible for chromosome movements occur in the chromosomes of all protozoan species so far described. They were depicted by L.R. Cleveland (1949) during mitosis in *Holomastigotoides tusitala*.

4) Telomeres. The chromosomes are also locked at their ends by specific DNA segments. These have been sequenced in protozoa and in a large number of species including humans. The very end of the eukaryotic telomere is composed of a tandem repeat

array of a short sequence. In several protozoa the terminal repeat sequence is TTTTGGGG and in many metazoa, including humans, it is TTAGGG (D. Kipling 1995).

5) The nucleolus is a most distinct structure which, in the light and electron microscopes, is identical in both protozoa and metazoa (G.F. Leedale 1968, G. Moyne et al. 1975, R.C. Brusca and G.J. Brusca 1990).

6) The nuclear envelope contains pores which are essentially alike in both types of organisms (E.W. Daniels et al. 1969, T.K. Golder 1976).

7) Microtubules are part of the spindle and are attached to centromeres in proto- and metazoa (L.R. Cleveland 1938, 1958, B. Bowers and E.D. Korn 1968).

8) Mitotic stages follow similar pathways but in protozoa there are variations of this process (C. Besse and J. Schrevel 1972).

9) Meiosis follows in the protozoa *Actinophrys* a sequence which is impressively similar to that of any higher organism including humans (K. Belar 1922).

10) Synaptonemal complex. As the chromosomes pair during meiotic prophase they build synaptonemal complexes that in protozoa are indistinguishable from those of other eukaryotes (F.O. Perkins and J.P. Amon 1969, N.V. Vinnikova 1976).

11) Polyteny. The similarity is striking in both the light and electron microscopes. The banding pattern found in the polytene chromosomes of *Stylonychia lemnae* cannot be easily distinguished from that of Dipteran species (D. Ammermann 1986). In human chromosomes polyteny also occurs but in a lower degree, in the form of endoreduplication (A. Levan and T.S. Hauschka 1953).

12) Spiral structure. The spiral coiling of the chromatids of *Holomastigotoides* and of *Amoeba proteus* (L.R. Cleveland 1949, I. Minassian and L.G.E. Bell 1976) is as evident as that of human chromosomes (Y. Ohnuki 1968).

13) Polyploidy. Chromosome numbers in protozoa vary extensively. Several species have only 2 chromosomes whereas others possess 600, such as *Amoeba proteus* (R.R. Kudo 1971). Triploidy (69 chromosomes) has been described in humans (J.A. Böök and B. Santesson 1960) and W. Engel et al. (1975) came to the conclusion that the human genome is of tetraploid origin.

14) Gene amplification occurs in protozoa as well as in higher organisms (M.-C. Yao 1986).

15) The occurrence of introns, as well as of RNA splicing, are also features of the protozoan chromosome (T.R. Cech et al. 1983, T.R. Cech 1986).

16) Structural changes may assume unexpected forms, as in the hypotrichous ciliates, in which the chromosomes undergo excision and elimination of DNA segments (L.A. Klobutcher and D.M Prescott 1986). Structural changes of many types occur regularly in the human genome (A.A. Sandberg 1980, S. Heim and F. Mitelman 1987).

17) Protozoa contain in their cytoplasm both chloroplasts (as in green plants) (R.C. Brusca and G.J. Brusca 1990, *Euglena*, G.F. Leedale 1967, *Euglena*) and mitochondria (J.P. Turner 1940). This means that their nuclear chromosomes also coexist with the DNA of other organelles.

18) DNA replication, RNA transcription and translation, resulting in protein synthesis, follow the same pathways and use the same type of molecules.

19) The genetic code is also the same in protozoa and in humans.

The inescapable conclusion is that the metazoan chromosome displays the same basic innovations that the chromosome of a protozoa has been able to acquire, or had initially acquired. This has occurred without the two types of chromosomes having had the opportunity to communicate genetically with each other, during the astounding period of time that separates them. Some protozoans are parasites of humans and other animals but most species are not. Moreover, parasitism does not necessarily imply genetic interchange. In addition, protozoan species have been exposed during this long evolutionary time to the most diverse environments.

The paradoxical situation is that the chromosome had no need of the long metazoan evolution, into invertebrates, plants and vertebrates, to produce all the mechanisms that it has invented. This also seems to imply that metazoan evolution has only followed what the evolution of the protozoan chromosome permitted and dictated.

It should be pointed out that some of the eukaryotic chromosome properties already existed in the bacterial chromosome, but they were only either a few or occurred in a rudimentary form. Obviously the properties of the eukaryotic chromosome could not come from nowhere. They had to be anchored on those of its predecessor. The bacterial chromosome is in a way an embryonic condition of the eukaryotic chromosome.

The research on the protozoan chromosome got its start in the 1920s with the pioneering work of K. Belar (1922). During the 1940s and 1950s it got a new impetus thanks mainly to the contributions of L.R. Cleveland (1949, 1954). But it was mainly in the 1970s and 1980s, that the use of electron microscopy combined with molecular methods, allowed a real understanding of the chromosome events and biochemical processes. It was then that the properties of the protozoan chromosome became well established.

MAINTENANCE OF ORGANIZATION.
THE PROTOZOAN VERSUS THE HUMAN CHROMOSOME -- 1

CLEVELAND 1938 - Mitosis in the protozoan *Barbulanympha*. Long centrioles form a central spindle and extranuclear chromosomal fibers connect with the intranuclear chromosomal fibers at the nuclear membrane.

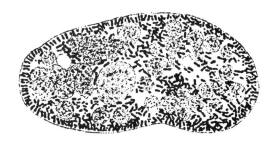

TURNER 1940 - Mitochondria in *Tillina canalifera*.

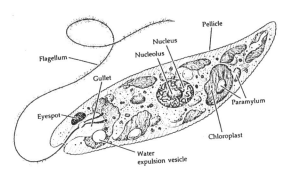

BRUSCA and BRUSCA 1990 - Anatomy of *Euglena* showing the nucleus, nucleolus and chloroplast.

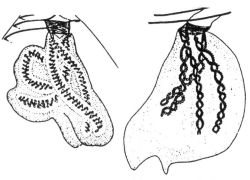

CLEVELAND 1949 - Prophase chromosomes in the protozoan *Holomastigotoides tusitala* showing major and minor spirals in the separated chromatids. The centromeres are attached to the rectangular spindle.

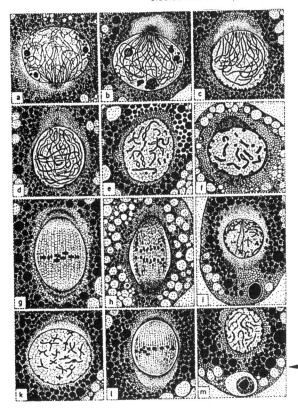

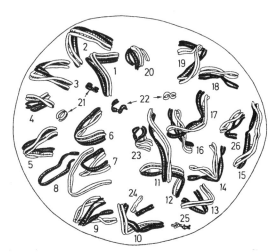

CLEVELAND 1954 - Meiosis in *Barbulanympha ufalula*. Diakinesis with 26 four-chromatid bivalents (numbered).

BELAR 1922 - The first and second divisions of meiosis in the protozoan *Actinophrys*.

MAINTENANCE OF ORGANIZATION.
THE PROTOZOAN VERSUS THE HUMAN CHROMOSOME -- 2

A

Histone	Residues	Molecular weight	NH₂ terminus
H1	163[a]	17,627[a]	Ala[b]
H2A.1	137[c]	14,654[c]	Acetyl-Ser[d]
H2A.2	132[c]	14,126[c]	Acetyl-Ser[d]
H2B	120[e]	13,433[e]	N-Trimethylalanine[f]
H3.1	135[g]	15,336[g]	Ala[g,h]
H3.2	135[g]	15,424[g]	Ala[g,h]
H4.1	102[i,j]	11,228[i,j]	Ala[i,j,k]
H4.2	102[i,l]	11,228[i,l]	Ala[i,l]

B

	H4				H3				H2A				H2B			
	COW	PEA	YEAST	TETRAHYMENA	COW	PEA	YEAST	TETRAHYMENA	COW	PEA	YEAST	TETRAHYMENA	COW	PEA	YEAST	TETRAHYMENA
COW	•	•	•	•	•	•	•	•	•	•	•	•	•	•	•	•
PEA	2	•	•	•	3	•	•	•	—	•	•	•	—	•	•	•
YEAST	8	9	•	•	11	12	•	•	15	—	•	•	19	—	•	•
TETRAHYMENA	20-22	19-21	23-25	•	14-18	14-19	19-21	•	25	—	25	•	31	—	36	•

GOROVSKY 1986 - A. Properties of the major histones of *Tetrahymena* macronuclei. B. Percentage differences between core histones of *Tetrahymena*, yeast, pea and cow.

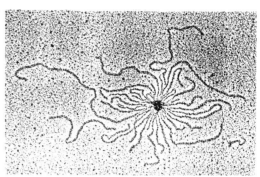

MURTI and PRESCOTT 1986 - Chromatin from the macronucleus of *Holosticha* sp. in the form of DNA molecules building rosettes.

OHNUKI 1968 - Spiral structure in human chromosomes.

MINASSIAN and BELL 1976 - Spiral structure in a nucleus of *Amoeba proteus*.

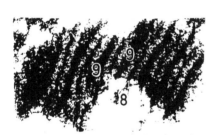

VINNIKOVA 1976 - Multiple synaptonemal complexes in *Dileptus*.

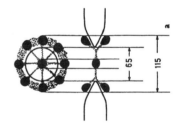

DANIELS et al. 1969 - Nuclear pore dimensions in *Amoeba* (in nm).

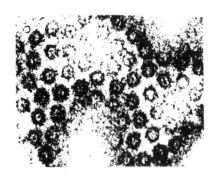

GOLDER 1976 - Electron micrograph of nuclear envelope of *Woodruffia*.

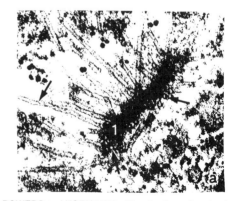

BOWERS and KORN 1968 - Ultrastructure of a microtubule organizing center in *Acanthamoeba castellanii*.

MAINTENANCE OF ORGANIZATION.
THE PROTOZOAN VERSUS THE HUMAN CHROMOSOME -- 3

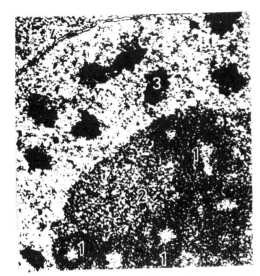

MOYNE et al. 1975 - Electron micrograph of the nucleolus of *Euglena gracilis* (1 and 2), with chromosomes (3) and nuclear envelope (4).

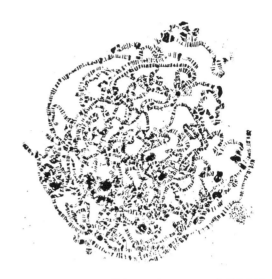

AMMERMANN 1986 - Polytene chromosomes in the macronucleus of *Stylonychia lemnae*.

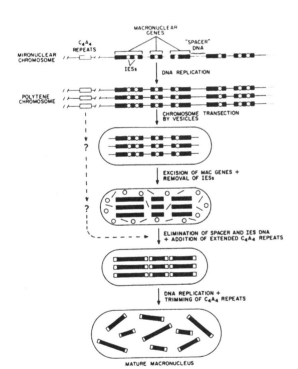

KLOBUTCHER and PRESCOTT 1986 - Summary of the developmental events transforming the chromosomal micronuclear genome into a macronucleus containing gene-sized DNA molecules in hypotrichous ciliates.

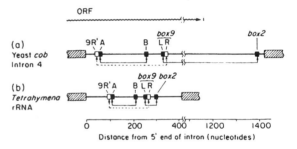

CECH et al. 1983 - Organization of conserved intron sequences (solid boxes) in *Tetrahymena*.

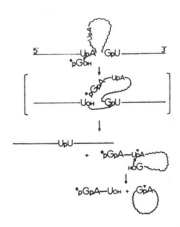

CECH 1986 - Transesterification mechanism for *Tetrahymena* pre-ribosomal RNA splicing.

Maintenance of the chromosome phenotype

Chromosome genotype and chromosome phenotype
The pattern of a chromosome is a phenomenon distinct from its genetic content. We tend to equate a chromosome with its DNA, but that is a simplification that results from the ignorance of the organization of the chromosome as a whole. Like a living organism, or a cell, every chromosome has a phenotype and a genotype. When inspected under the light, or the electron microscope, the observer is not seeing the chromosome genotype, which consists of its DNA, but is looking at an edifice built by a series of other molecular components. The chromosome phenotype is built mainly by: (1) Histone proteins which are intimately associated with the DNA in the form of nucleosomes. (2) Structural proteins, which build the chromosome scaffold. (3) Transcription factors, which are a third class of proteins, that regulate transcription. (4) High-mobility proteins which are necessary for cell viability. (5) The various types of RNA molecules that are part of the chromosome. As shown by A. Klug (1985) in a condensed chromatin fiber each octameric histone core is a disk and each nucleosome associates with one H1 (histone) molecule. The fiber coils in a solenoid with a diameter of 30 nm.

It is this enormous molecular assembly that we are facing when we study a prophase, metaphase or an anaphase chromosome under the microscope. Its phenotype changes dramatically during these stages, from a nearly invisible thread at prophase, to a thick rod at anaphase.

The significance of the structural proteins, in building the chromosome phenotype, already became evident when A.E. Mirsky and A.W. Pollister (1946) removed, with the help of enzymes, both the DNA and the histones of metaphase chromosomes, without altering their basic structure. After such a treatment the chromosomes looked nearly intact. At present it is known that this striking conservation of the chromosome's form and structure is due to the regular helical pattern of the scaffold proteins (J.R. Paulson and U.K. Laemmli 1977, A.J.F. Griffiths et al. 1993). All these macromolecules have been conserved throughout the eukaryotic evolution, a feature that is at the basis of the maintenance of the same chromosome pattern in plants and animals, including humans.

Prediction of pattern display
We tend to forget that one of the best examples of chromosome rigidity is found in the repetitive expression of its phenotype. Every mitosis starts with a prophase which is followed by a metaphase, that leads to an anaphase and finishes in a telophase. Remarkable, is that the chromosomes display a phenotype which is characteristic of every phase. This is due to a sequence of spiralization and despiralization that is strictly canalized and that has been maintained unaltered, in its main features, for over a billion years.

Meiosis is a variant of mitosis, and in this cellular transformation the same degree of chromosome rigidity is displayed. A typical example is pachytene. Every cytologist knows in advance, that when inspecting this cell stage – be it in plants or animals – a specific chromomere pattern is always present. The same constancy holds for other meiotic stages.

The rigidity of chromosome organization is so extreme that it is possible to predict what kind of basic chromosome pattern is exhibited at every stage of cell division. However, since the chromosome is the product of a genetic constitution that varies from chromosome to chromosome, this conservation of the basic pattern, takes different forms, in different organisms, and in individual chromosomes. Yet the underlying order surfaces in the form of essentially identical solutions.

Let us consider some of the salient features of this behaviour.

Heterochromatin – This is a term coined by E. Heitz (1932, 1934) when he found that specific regions of the chromosomes remained visible during interphase and appeared highly coiled at other stages. He also noticed that the heterochromatic regions did not occur at random but tended to occur at specific sites. These heavily stained regions turned out to consist of large chromomeres. Heterochromatin was later found to consist of segments poor in structural genes and in which DNA satellites are present (P.M.B. Walker 1971). This is the classic example of the rigidity of pattern organization. In plants and animals, including humans, it has been observed to usually occur near centromeres and at the ends of the arms (V. Gregoire 1932, E. Heitz 1933, S. Pathak 1979, T. Schmidt and J.S. Heslop-Harrison 1998).

Chromomere gradient – The chromomere is an individualized region of the chromosome which takes the form of a stained spherical structure due to tight coiling of the chromatids. This phenotype is mostly distinct during the prophase of mitosis and meiosis. It was extensively studied by J. Belling, and others, in the 1920s and 1930s. Chromomeres, like the bands of polytene chromosomes, are present along most of the chromosome body. It was found that in many plant species (belonging to quite different families), and in certain animals, the chromomeres are large on both sides of the centromere and decrease successively in size towards the telomeres building a gradient. The degree of decrease was found to be dependent on the distance telomere-centromere (A. Lima-de-Faria 1954, P. Eberle 1956, see page 176).

Knobs – This is a term introduced by maize geneticists to denote a heavily stained group of chromomeres. They vary in position in different cultivated forms of maize but in plants from many species they tend to occupy a terminal or subterminal position (A.E. Longley 1938).

Constancy of distance between chromomeres – *Ornithogalum virens* is a species with only 3 chromosomes (haploid) each of them being easily distinguished from the others. At pachytene the three chromosomes have a total chromomere number of 274.2, but surprisingly the same chromosomes, at the second prophase of the pollen mitosis display only 67.4 chromomeres. As the chromosomes shorten at this last stage the chromomeres do not come together, building a more compact chromosome, but maintain their average distance. This is also the case in other plant and animal species. The average chromomere distance turned out to be 0.95 microns in plant species belonging to different families irrespective of the degree of contraction of the chromosome. So far there is no molecular explanation for this ordered coiling of the chromosome (A. Lima-de-Faria et al. 1959).

Cold-induced regions – L.F. La Cour (1951) studied extensively these regions in *Trillium*. Following cold treatment of the plants, less stained and thinner chromosome segments, appeared in the metaphase chromosomes of many species. They also turned out to have an ordered distribution and to occur mainly near centromeres in both plants and animals (such as *Amblystoma mexicanum*). The relationship of these less stained regions with heavily stained heterochromatin is not understood.

Chiasmata – Chiasmata are the regions that hold together the chromosomes of a bivalent at metaphase I of meiosis. Their location was extensively studied in the 1930s and 1940s. It also turned out, that in a large group of both animals and plants, the chiasmata were formed at fixed positions. These locations tended to be related to centromere position in *Paris* and in *Fritillaria* (C.D. Darlington 1941).

All these structural features of the chromosome phenotype have been maintained practically intact since the dawn of eukaryotic evolution. The question arises: What molecular mechanisms are responsible for these phenomena?

MAINTENANCE OF THE CHROMOSOME PHENOTYPE -- 1

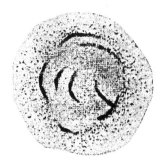

GREGOIRE 1932 - Nucleus of *Impatiens* showing heterochromatin near centromeres.

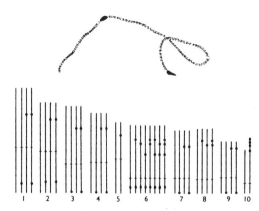

LONGLEY 1938 - Knobs of maize chromosomes as seen at pachytene. Above, chromosome 3 showing one terminal and one interstitial knob. Below, diagrams of the 10 maize chromosomes with positions of knobs from different maize varieties.

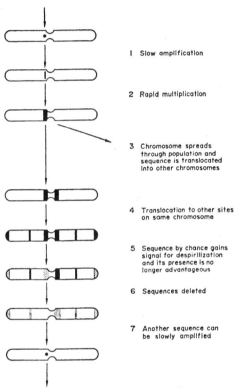

1 Slow amplification
2 Rapid multiplication
3 Chromosome spreads through population and sequence is translocated into other chromosomes
4 Translocation to other sites on same chromosome
5 Sequence by chance gains signal for despirilization and its presence is no longer advantageous
6 Sequences deleted
7 Another sequence can be slowly amplified

WALKER 1971 - Life history of a satellite DNA: the whole cycle.

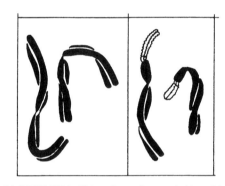

LA COUR 1951 - Heterochromatin revealed by cold treatment in *Trillium*.

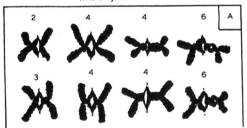

DARLINGTON 1941 - Distribution of chiasmata in bivalent chromosomes of *Paris*. Note the repeated occurrence of chiasmata on both sides of the centromere and at equal distances.

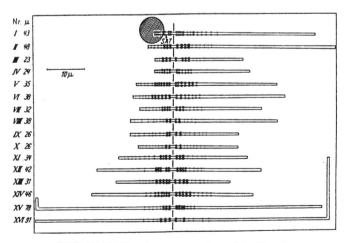

EBERLE 1956 The chromomere pattern of the 16 pachytene chromosomes of *Aeschynanthus*. The vertical line marks the position of the centromere.

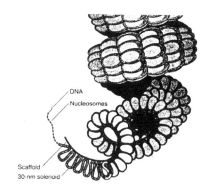

LIMA-DE-FARIA 1954 -
Homologous chromosomes of *Agapanthus umbellatus* at pachytene (top), prophase II of meiosis (middle) and prophase of mitosis (bottom) showing chromomere gradient starting on both sides of centromere. The same pattern and chromomere distance are maintained independently of the large difference in chromosome length.

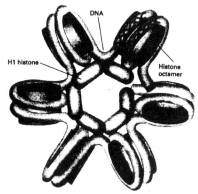

KLUG 1985 - View of a model of the condensed chromatin fiber. The octameric histone core is shown as a disk. Each nucleosome associates with one H1 (histone) molecule. The fiber coils into a solenoid with a diameter of 30 nm.

GRIFFITHS et al. 1993 - Model of the association of DNA, histones and scaffold proteins in a metaphase chromosome. The chromosome scaffold coils into a helix. The DNA and nucleosomes build a 30 nm solenoid.

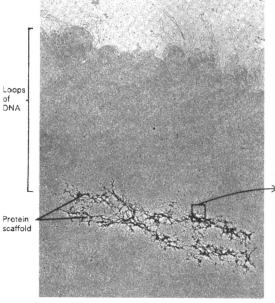

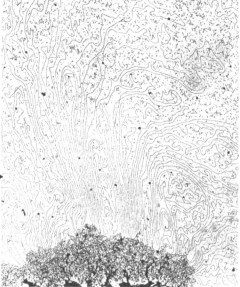

PAULSON and LAEMMLI 1977 -
Electron micrograph of a histone-depleted metaphase chromosome prepared from human cells by treatment with a mild detergent. Left, lower magnification showing the nonhistone protein scaffolding (dark structure) which retains the shape of the metaphase chromosome. Loops of DNA protrude from the protein structure. Right, higher magnification showing the apparent attachment of the DNA loops to the protein scaffold.

Maintenance of gene order

Geneticists and cytologists, in the 1950s and 1960s, tended to think of the chromosome as a disordered body submitted to mechanical forces, in which crossing over was the main agent of its disruption. The location of genes was supposed to be constantly altered and long-term mutation changed the genes beyond recognition.

Embryologists, who compared the organ and tissue transformations, occurring during the embryogenesis of distantly related animals, had difficulties in accepting this explanation. Development showed such a regularity, that this order ought to be anchored on the chromosomes. They were, however, silenced by the orthodoxy of geneticists who would not allow another interpretation.

Suddenly, as molecular biology came of age the opposite turned out to be true. In the early 1970s, as the molecular techniques used in bacteria were employed in higher organisms, a surprising fact became evident. Genes had not changed appreciably for billions of years. Not only were certain genes the same in all eukaryotes but this conservatism also extended to the prokaryotes.

Order of genes in bacteria and viruses
The genetics of bacteria opened the way to the study of the exact location of genes and of their functional interactions. The concept of the operon, developed by F. Jacob and J. Monod (1960) was a breakthrough in this field. The lactose operon in *E. coli* revealed three new basic features: 1) Genes with related functions were clustered, and were contiguous, in a specific region of the chromosome. 2) The transcription of the genes was polarized, *i.e.* there was a sequence in the synthesis of the messenger RNA. 3) The order of the genes along the chromosome was related to their metabolic role in the cell. Other operons, in *E. coli* and in other bacterial species, were soon found to display the same basic features. It also turned out that, as in the case of the tryptophan operon, some genes associated in separated clusters due to translocations (I.P. Crawford 1975). In viruses, such as lambda, the genes which were functionally related were also found to build clusters (R.W. Old and S.B. Primrose 1980).

Clustering of genes with similar functions in higher organisms
The clustering of genes with similar functions was subsequently found to be of general occurrence. Examples are the mimetic mutants of tomato which are involved in leaf development (C.M. Rick 1971) and the genes of the mouse responsible for tail formation (M.C. Green et al. 1973). The human chromosome No. 9 showed clustering of housekeeping genes and No. 11 had a cluster comprising more than nine oncogenes. Polarity of gene transcription also occurred in human chromosomes. The classical example of gene clustering has been, for many years, the human X chromosome in which genes with similar functions are located (V.A. McKusick 1971).

Gene conservation
Genes not only tend to be clustered and polarized but they have been conserved throughout millions of years. M.O. Dayhoff (1969, 1972) was a pioneer of this study. Protein and nucleic acid sequences provided evidence that the genes for cytochrome c, ferredoxin, trypsin, transfer RNA and 5S ribosomal RNA had been highly conserved during the evolution from bacteria to humans. For instance, for cytochrome c about 60% of the total number of amino acids at corresponding positions were identical in wheat and human chains, while 30% were identical in humans and *R. rubrum* (a bacterium). Among gene relics are the pseudogenes. Five pseudogenes are present in the human globin gene families (E.F. Fritsch et al. 1980). Since then many other types of genes have been found to be conserved such as those involved in cell death pathways and in chromatin silencing (E.H.-Y.A. Cheng et al. 2001, D. Moazed 2001).

Maintenance of order within the structure of the gene
Since then, a plethora of genes has been sequenced. Moreover their intron and exon numbers, as well as their flanking sequences, have been determined. The genes of maize and of a chicken can be aligned, as well as those of an insect and of a human, showing great similarity. Their identity can be determined with all the accuracy furnished by the base analysis. Small differences that have emerged, during the large period of time separating these species, can also be accurately determined. This is exemplified by the gene for the enzyme triose phosphate isomerase when its organization in the chicken is compared with that in maize (L.A. Fothergill-Gilmore 1991). This same gene was used in the study of the origin of split genes. Its evolution was traced from bacteria to vertebrates disclosing when the gene was split into exons and introns (W. Gilbert et al. 1987).

Recently it has been found that humans and mice share about 99% of their genes (J.L. Goldstein 2001 and T.J. Hudson et al. 2001).

Maintenance of gene order along the chromosome body

One of the most striking cases of this phenomenon is the homeobox gene family. Until only a few years ago the wings of insects were considered to be unrelated to those of birds and no one would dare to compare the body segmentation of a worm, or the body pattern of a plant, with that of a human. At present it is known that the same homeotic genes are responsible for the segmentation of the body of a worm, of an insect, of the sequence of the different parts of a flower and of the segmentation of the human vertebral column (M.P. Scott 1992). Hence, gene permanence has been extreme.

But the conservatism has also involved other features: Within a gene complex its various components have maintained their order during millions of years of evolution. As B. Alberts et al. (1994) point out, one of the "deep mysteries" of the homeobox domain is how it operates and how it is organized. As they state: "the sequence in which the genes are ordered along the chromosome in both the Antennapedia and the bithorax complexes, corresponds almost exactly to the order in which they are expressed along the axis of the body" (B. Alberts et al. 1994). But still more remarkable is that this feature of the gene complex has been highly conserved in the course of evolution being the same in flies and in humans (H. Warmus and R.A. Weinberg 1993).

Self-assembly may be responsible for molecular order

Of significance for understanding the mechanism that may have been responsible for the emergence of gene order and conservation, are the recent data on the origin and evolution of the genetic code. From being considered to be a product of random events it is now seen as the result of ordered processes of atomic self-assembly. The code might have originated from natural interactions between RNA molecules and amino acids, and the same codon reallocations may have occurred more than once (R. Cedergren and P. Miramontes 1996, R.D. Knight et al. 1999).

MAINTENANCE OF GENE ORDER -- 1

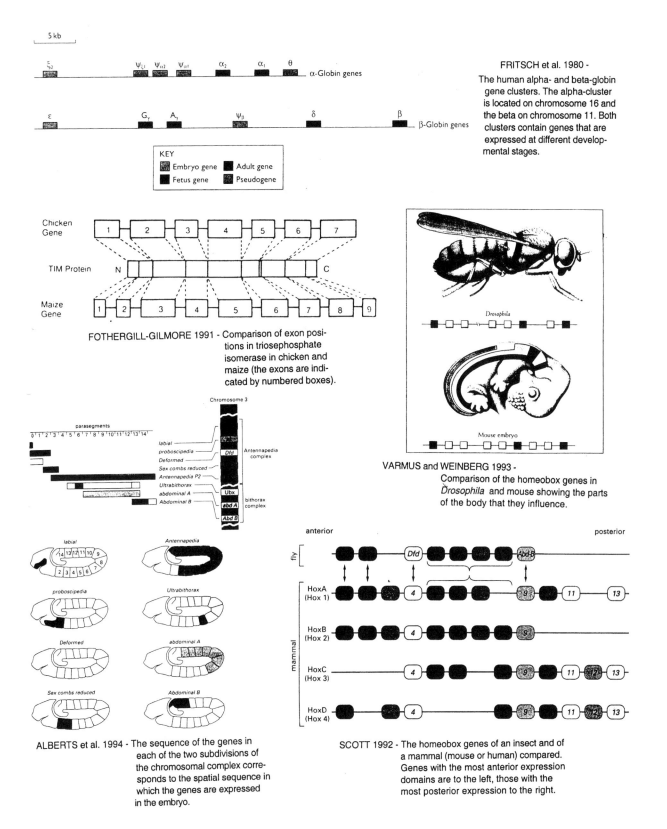

FRITSCH et al. 1980 - The human alpha- and beta-globin gene clusters. The alpha-cluster is located on chromosome 16 and the beta on chromosome 11. Both clusters contain genes that are expressed at different developmental stages.

FOTHERGILL-GILMORE 1991 - Comparison of exon positions in triosephosphate isomerase in chicken and maize (the exons are indicated by numbered boxes).

VARMUS and WEINBERG 1993 - Comparison of the homeobox genes in *Drosophila* and mouse showing the parts of the body that they influence.

ALBERTS et al. 1994 - The sequence of the genes in each of the two subdivisions of the chromosomal complex corresponds to the spatial sequence in which the genes are expressed in the embryo.

SCOTT 1992 - The homeobox genes of an insect and of a mammal (mouse or human) compared. Genes with the most anterior expression domains are to the left, those with the most posterior expression to the right.

MAINTENANCE OF GENE ORDER -- 2

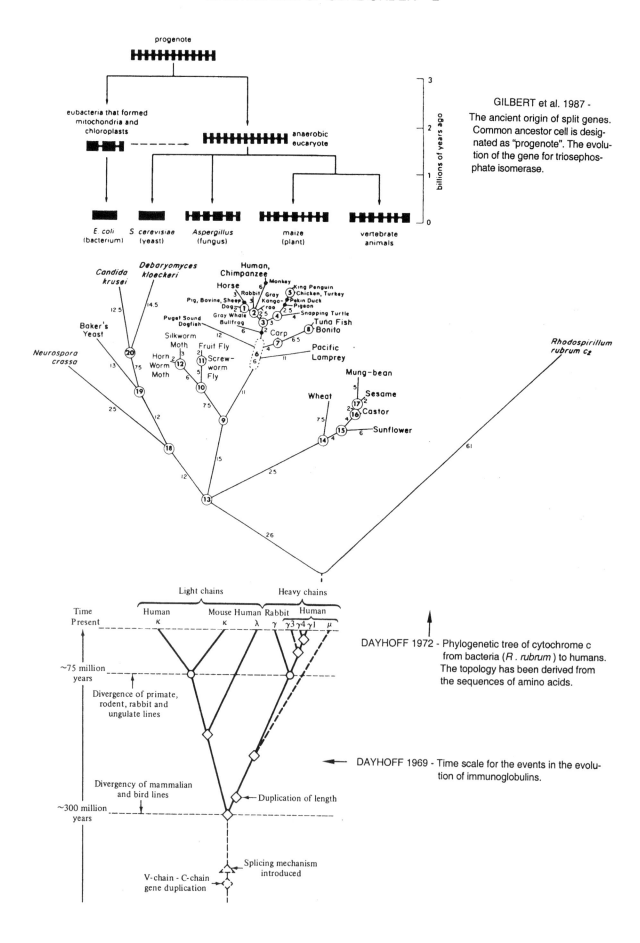

GILBERT et al. 1987 - The ancient origin of split genes. Common ancestor cell is designated as "progenote". The evolution of the gene for triosephosphate isomerase.

DAYHOFF 1972 - Phylogenetic tree of cytochrome c from bacteria (*R. rubrum*) to humans. The topology has been derived from the sequences of amino acids.

DAYHOFF 1969 - Time scale for the events in the evolution of immunoglobulins.

Maintenance of function

One of the main contributions of molecular cytogenetics has been the demonstration that not only the structure, but also the function of chromosomes is so rigid, that it has been preserved for millions of years. This is evidenced by several main features such as: (1) The maintenance of the function, in a large number of structural genes, since the emergence of prokaryotes. (2) The transfer of these genes in an ordered way from species to species in the form of chromosome blocks, translocated in a nearly intact state (syntenic genes). In many cases this has taken the form of whole arm transfer.

Several mechanisms have contributed to preserving the permanence of function. (1) DNA repair, a process that corrects spontaneous damage occurring in the DNA molecule and that results in the preservation of the original function. (2) DNA proofreading. (3) The monitoring of messenger RNAs for errors, called "RNA surveillance", which results in corrected products of gene expression. (4) Other mechanisms may have been involved which have ensured the conservation of the promoter and operator sequences. (5) There may also have been, molecular mechanisms that have maintained the integrity of the gene sequence itself.

Let us consider these processes.

The stability of most structural genes
Coenzymes are of universal occurrence in all living organisms and there has been no or little change in the structure of coenzymes throughout evolution. Other proteins have also been conserved to a high degree. One of them is histone-IV a component of the nucleosome. The evolution of the sequences of proteins led M.O. Dayhoff et al. (1975) to conclude that all proteins known may belong to fewer than 1000 distinct superfamilies (groups related to their amino acid sequences). Cytochrome c, haemoglobin, and chlorophyll, are macromolecules which are of general occurrence. Moreover, the DNA sequences of their genes have varied so little that the final molecule has the same basic function. Even the haemoglobins found in plants have an amino acid sequence not very different from that of animals. Hence mechanisms, not yet fully understood, may have been at work, maintaining the integrity of the promoter, of the operator and of the DNA sequences that are responsible for the protein formation.

Gene synteny – chromosomal segments that are homologous between species
Since the early days of cytology it has been known that whole arms are transferred between chromosomes, by means of translocations. This is most common in animal species and has led to the preservation of gene function by moving a large group of genes without altering their mutual interactions.

The combination of *in situ* hybridization with fluorescent chromosome probes (molecular clones belonging to a specific chromosome) has allowed demonstrating the existence of conserved gene segments in chromosomes from different species of mammals. This has provided information on the composition of ancestral chromosomes by comparing gene blocks of human (HSA), pig (SSC), cattle (BTA), Indian muntjac (MMV), horse (ECA), harbor seal (PVI), cat (FCA), American mink (MVI) and mouse (MMU) genomes (B.P. Chowdhary et al. 1998). Another study mapped 96 human loci in the chromosomes of the cat (human chromosomes showing banding are compared with cat chromosomes in solid black) (S.J. O'Brien et al. 1997).

DNA repair and proofreading
As B. Alberts et al. (1994) have pointed out "If left uncorrected, spontaneous DNA damage would rapidly change DNA sequences". Thermal collisions with other molecules would lead to significant gene changes. It is known that about 5,000 purine bases are lost per day from the DNA of each human cell because of thermal disruption of certain atoms. This and other accidental base changes in the DNA molecule are eliminated by DNA repair. The altered portion of DNA is recognized and removed by repair nucleases. Then DNA polymerase copies the correct strand and this is sealed by a DNA ligase. Other enzyme pathways may be involved in the restoration of the original function. DNA polymerases also participate in the proofreading mechanism that ensures that DNA replication occurs with a minimum of mistakes. The fidelity of copying is of the order of one error in every 10^9 base pair replications. A protein, called p53, that exerts a tumor-suppressing function, has been found to facilitate DNA repair (H. Tanaka et al. 2000, G. Lozano and S.J. Elledge 2000).

RNA surveillance
M.R. Culbertson (1999) points out that the role of RNA surveillance has unforeseen consequences for gene expression in cancer and other diseases. In this process the messenger RNAs are monitored for errors that arise during gene expression. Base substitutions often cause chain termination. This results in the building of nonsense RNAs, but these rarely produce truncated proteins. These RNAs, which code for protein fragments, are recognized and eliminated.

This RNA surveillance has been found to occur in fungi, plants, nematodes and vertebrates. This is another impressive molecular mechanism that enables the chromosome to maintain its functional rigidity.

It seems reasonable to assume that DNA repair, DNA proofreading and RNA surveillance may have been sufficient to safeguard the continuity of function. But so many new, and unexpected, mechanisms have turned out to be part of the chromosome organization, that we should not be surprised if, in the future, additional molecular processes will contribute to better elucidate the permanence of function.

MAINTENANCE OF FUNCTION.

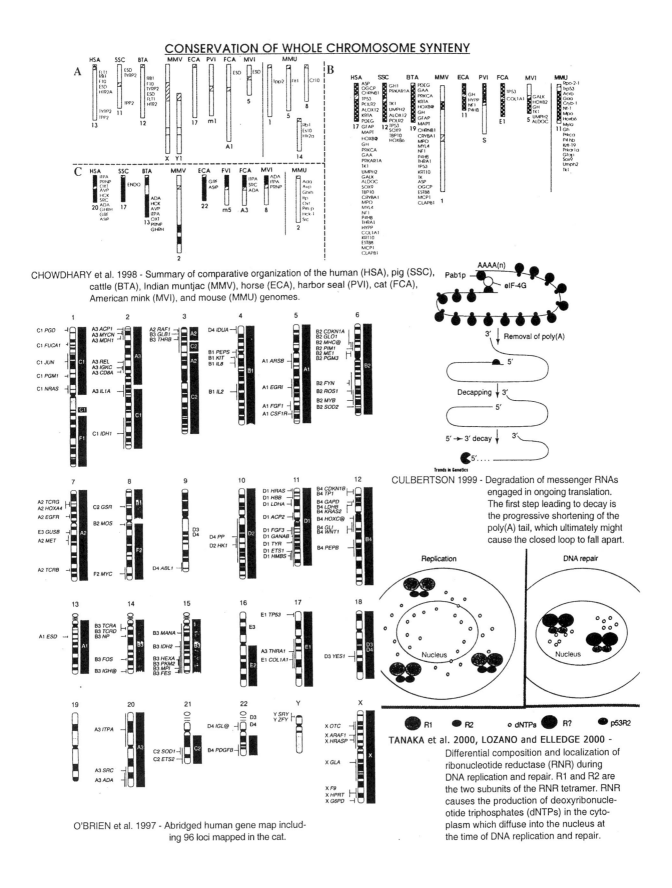

The periodicity of chromosome transformations

Convergence in evolution was recognized long ago. It has been considered to be an accidental phenomenon mostly related to the repetition of similar environmental conditions faced by organisms. For this reason convergence has been ascribed a marginal role in evolution (G.R. Taylor 1983).

But the repetition of the same macromolecular solution in the chromosome is a quite different phenomenon: 1) It is a modification not occurring in the organism's phenotype but in a cellular organelle with its own molecular organization. 2) It appears in species such as plants, insects and mammals which are not directly related phylogenetically. 3) It is not necessarily correlated with the environment since these organisms have habitats which are quite disparate.

Chromosome evolution has led to the periodic occurrence of several identical solutions in its structure and organization. Some of these are: 1) The production of extremely low, as well as extremely high, chromosome numbers in phylogenetically unrelated organisms. 2) The emergence of multiple translocations, leading to complex types of rearrangements, in both plants and animals. 3) The display of large heterochromatic regions, on both sides of the centromere, in highly diversified plant families and in animals.

The occurrence of very low chromosome numbers in plants
Geneticists in the 1920s and 1930s were puzzled by the fact that several plant species had only 3 chromosomes, as their haploid number. How could, the whole genetic information to make a flowering plant, be packed in only 3 chromosomes? Later the situation became even more extreme. *Haplopappus gracilis* was found to have only 2 chromosomes (haploid) (R.C. Jackson 1959). Some plants may even be chimaeras with tissues having the normal diploid number of 4 chromosomes, but with other tissues surviving equally well with only the haploid number of 2 chromosomes (A. Lima-de-Faria and H. Jaworska 1964). Soon another plant species *Brachycome lineariloba* was discovered with n = 2 (C.R. Carter and S. Smith-White 1972). What subsequently became evident was that within the genus *Haplopappus* there were also species with 2n = 6, 8, 10, 12, 18, 24 and 36 chromosomes. This meant that a plant with practically the same structural and functional features (since it was recognized as belonging to the same genus) could be produced independently of whether its genes had been compressed into 4 or dispersed into 36 chromosomes. It was not only the gene number, but the degree of gene expression, that had to be under control, to lead to the production of a nearly identical plant phenotype.

The occurrence of very low chromosome numbers in insects
Insects are not directly related to flowering plants. These invertebrates emerged about 450 million years ago, whereas flowering plants appeared only 181 million years ago (R.D. Barnes 1980). Yet they show the same extreme low chromosome numbers. The coleoptera *Chalcolepidius* has only 2 chromosomes (haploid) (A. Ferreira et al. 1984). The same number is found in the Diptera *Cricotopus silvestris* (P. Michailova 1976).

The ants have been studied extensively. Almost 500 species of this insect group have been karyotyped. The ant *Myrmecia pilosula* has closely related species with 2n = 9, 10, 16, 24, 30, 31 and 32 chromosomes, but siblings "currently indistinguishable morphologically" from these have turned out to have only 2 chromosomes in the females and one single chromosome in the males of this species (M.W.J. Crosland and R.H. Crozier 1986). Like in honey bees, the females develop from fertilized eggs, but the males are produced without the eggs being fertilized, the result being that they have only half of the chromosomes, which in this case is a single one. According to H.T. Imai et al. (1990), who has made the most extensive study of this insect group, there are ants that contain as many as 94 chromosomes (diploid). Hence the variation is 2 to 94, and yet the final genetic product is still an ant.

Low chromosome numbers also occur in protozoa
Protozoa are unicellular eukaryotes but their chromosomes have most of the features of those of metazoa. According to the fossil record they emerged 1,600 million years ago. They also display an astonishing variation in chromosome number. *Spirotrichonympha polygyra* has only 2 chromosomes (diploid) but many species have intermediate numbers such as 2n = 6, 8, 12, 14, 16, 20, 30, 40, 200 until they reach one of the highest values known, 500–600 chromosomes (diploid) in *Amoeba proteus* (R.R. Kudo 1971). In this case the same type of unicellular organism is produced using either 2 or 500 chromosomes.

But the low numbers also appear in higher mammals
Geneticists tolerated the view that protozoa, plants and insects could be produced with few chromosomes, but not the highly evolved vertebrates with

their complex tissues, especially mammals. It thus came as a surprise when a deer, *Muntiacus muntjak* was found to have only 6 chromosomes (diploid female) and 7 (diploid male) (D.H. Wurster and K. Benirschke 1970). Even more surprising was the finding that, a nearly identical species *Muntiacus reevesi* had the same chromosome number as humans 2n = 46 (T.C. Hsu and K. Benirschke 1971, D. Whitely and M. Nixon 1972) and that other deer species, such as the reindeer, had 2n = 70. The DNAs of these species were extracted and cleaved with restriction enzymes. DNA-DNA hybridization disclosed that the DNA of *M. muntjak* contained many sequences which were homologous to the other deer species which have a chromosome number 10 times higher (A. Lima-de-Faria et al. 1986).

In other mammals the chromosome number may also reach extreme heights. The water rat *Ichthyomys pitteri* has 2n = 92. Hence a most complex mammal can be produced independently of its genes being harboured in 3, 23, 35 and 46 chromosomes (haploid).

A feature which has been found to be characteristic of the chromosome variation in protozoa, plants, insects and mammals, is that as the chromosome number decreases, the size of the chromosomes tends to increase. This is clearly the case in the ants and the deer species. Moreover, in the case of *Muntiacus* it is known that: (1) the amount of DNA is about the same in the 2n = 6 and 2n = 46 chromosomes, (2) the small centromeres of the 46 chromosomes have been grouped in the large centromeres of the 3 chromosomes (B.R. Brinkley et al. 1984), (3) gene blocks were transferred as units between the small and large chromosomes, (4) both an increase and a decrease in chromosome number has occurred during species evolution.

A feature that is repeated in this type of periodicity is that both in *Haplopappus* and in the ant *Myrmecia* the diploid chromosome number can even be reduced to the haploid value producing the same type of organism. In the plant 4 chromosomes can be reduced to two, and in the insect, two can be reduced to one. This information gives a first hint of the mechanisms involved in this process of chromosome diversification.

What is significant from the point of view of rigidity is that the chromosome is able to produce the same solution leading to the periodic recurrence of the same genetic event. The chromosome disperses or condenses, again and again, its genetic content, irrespective of whether it is present in a simple organism, such as a protozoa, or in a most complex one, such as a higher mammal.

It is, at present, unknown what kind of molecular mechanism obliges the chromosome to exist in the cell as 2 or as 500 copies and yet be able to produce a basically identical phenotype.

Other forms of chromosome behaviour show periodicity

One of them is the complex system of reciprocal translocations that results in the formation of long rings, or chains, of bivalents at metaphase I of meiosis. These were studied extensively in *Oenothera* (R.E. Cleland 1936), *Rhoeo*, and other plant species, belonging to quite different families. *Oenothera* is an Onagraceae whereas *Rhoeo* is a Commelinaceae. The same chromosome solution also appeared in an animal, the crustacean *Mesocyclops edax* (C.C. Chinnappa and R. Victor 1979). Suddenly, and in a species in which it would least be expected, the same chromosomal behaviour takes place, establishing a periodic event.

Another phenomenon, is the sudden formation of large blocks of heterochromatin on both sides of the centromere. These occur in species belonging to at least 16 quite different plant families ranging from the Gramineae to the Balsaminaceae. They also appear in two distinct groups of animals: insects and mammals (several species including the mouse and humans). Again, the chromosome follows its independent molecular path, periodically exhibiting a given property irrespective of the complexity of the organism in which it happens to be present. Other examples of this intriguing property could be added.

Proteins have also been found to display periodicity. This extends the phenomenon to macromolecules acting at other levels of cell organization. Examples are the enzymes staphylococcal nuclease (J.W.R. Schwabe and A.A. Travers 1993) and carbonic anhydrase (A. Liljas and M. Laurberg 2000). Until recently it was believed that enzymes had to possess the same amino acid sequence to perform the same function. These proteins were found to have quite different amino acid sequences, and different tertiary and quaternary structures, and yet they produce the same enzymatic reaction. This same functional solution occurred in bacteria, plants and humans.

The protein periodicity may turn out to be a key phenomenon in the understanding of chromosome periodicity.

THE PERIODICITY OF CHROMOSOME TRANSFORMATIONS -- 1

MICHAILOVA 1976 -
The two polytene chromosomes of the Diptera *Cricotopus silvestris* in salivary glands and at mitosis (upper right) at the side of a 10 micron bar.

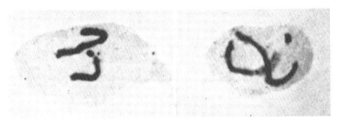

LIMA-DE-FARIA and JAWORSKA 1964 -
Two mitoses in haploid tissues of the plant *Haplopappus gracilis* which has only two chromosomes.

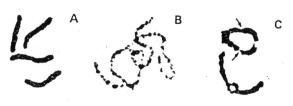

FERREIRA et al. 1984 - The 2 chromosomes of the Coleoptera *Chalcolepidius zonatus*. A, mitotic metaphase 2n=4. B, pachytene. C, diplotene (arrows indicate centromeres).

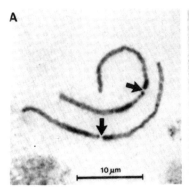

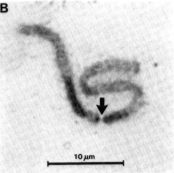

Myrmecia pilosula
workers 2n=2 males n=1

CROSLAND and CROZIER 1986 -
The chromosomes of the ant *Myrmecia pilosula*. The two metaphase chromosomes in a worker (A) and the single chromosome in a male. A single chromosome can produce a perfect animal.

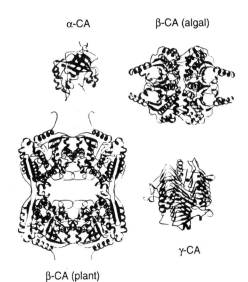

LILJAS and LAURBERG 2000 -
The tertiary structures of different forms of carbonic anhydrase (human, algal, plant and from archae). Their amino acid sequences are quite different yet they have the same enzymatic function.

THE PERIODICITY OF CHROMOSOME TRANSFORMATIONS --2

Muntiacus muntjak
n = 3

Muntiacus reevesi
n = 23

WHITELEY and NIXON 1972 -
Muntiacus muntjak, female n=3 (1A), male n=3+y_2 (1). *Muntiacus reevesi* n=23, female (2A), male (2). The females are practically identical despite the extreme variation in chromosome number.

CLELAND 1936 - Chromosome configurations in *Oenothera*. A ring of 12 chromosomes plus one pair at diakinesis and a diagram of the ring association.

Y_2 X Y_1
Male

X X
Female

WURSTER and BENIRSCHKE 1970 -
The mitotic chromosomes of *Muntiacus muntjak* (male 2n=7, female 2n=6).

CHINNAPPA and VICTOR 1979 -
The ring of 14 chromosomes in the crustacean *Mesocyclops edax* due to segmental interchanges as in *Oenothera*.

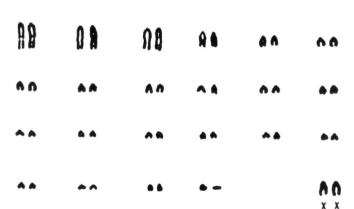

HSU and BENIRSCHKE 1971
The mitotic chromosomes of *Muntiacus reevesi* 2n=46 (female).

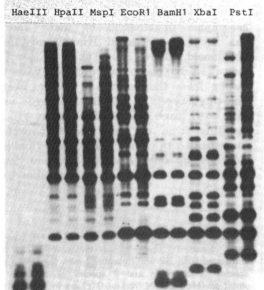

LIMA-DE-FARIA et al. 1986 - The DNA of *Muntiacus muntjak* (n = 3) was cleaved with 7 restriction enzymes (Hae III to Pst I) and hybridized with the DNA of red deer (n = 34) using a nick translated band of 770 base pairs. The DNA-DNA hybridization shows that *M. muntjak* contains in its DNA many sequences that are homologous to the deer species which has a chromosome number 10 times higher. In situ hybridization revealed that these sequences were located near centromeres in 32 of the 34 chromosomes.

THE CHROMOSOME'S PLASTICITY

Structural change

What kind of organelle is the chromosome? It remains a most viable, rigid and conservative body. Yet it is able to change its structure, pattern, size and number in minutes, better than any magician could do. Few structures embody the antithetical nature of life's construction, in such a patent way, as the eukaryotic chromosome.

From where arose all the molecular mechanisms responsible for these "magical tricks" that allow the chromosome: (1) to be broken but immediately resealed, (2) to have its segments inverted but continuing to function with equal efficiency, (3) to increase its size 100 fold maintaining its genetic constitution, (4) to have been restructured for over 2 billion years yet maintaining its organization, (5) to change within minutes its shape during cell division, and at the same time to move to the spindle poles with impressive efficiency. The list could be made longer.

Let us consider first the structural changes, or rearrangements, that the chromosome has gone through, during millions of years of organism evolution and during the thousands of cell divisions that occur when a fertilized egg develops into an adult organism. The chromosome has been, and is being, restructured at every moment of its cellular existence.

The structural changes are of many possible kinds. DNA segments may be translocated to other chromosomes, or within the same chromosome, other sequences may be deleted or added. The chromosome is actually a "Pandora's box" from which most kinds of different rearrangements can be extracted. These have been summarized by C.D. Darlington (1937). Other instances are: (1) the inversions studied largely in *Drosophila* (K. Pätau 1935), (2) the additions and subtractions of small groups of bands as in the case of the bar gene studied by C.B. Bridges (1936), and the deficiency-duplication cycle discovered in maize by B. McClintock (1938) that led to the formation of chromosome rings.

The rearrangement of the chromosome, is such a permanent event, that it occurs regularly at every meiotic division during the pairing that takes place at its first prophase. At this stage DNA segments are exchanged, a phenomenon that has been called crossing over. C.D. Darlington (1932) has, in a diagram, visualized its effects, in which gametes are produced with new DNA combinations. Crossing over may also result in the formation of bridges and fragments at anaphase (Th. Dobzhansky 1941). J. Belling (1931) was among the first to propose a mechanism for strand exchange during crossing over (M.J.D. White 1973). The cytological demonstration of this phenomenon was obtained by B. McClintock (1930) in maize and by C. Stern (1931) in *Drosophila* (summarized by L.W. Sharp 1934).

The ability to exchange segments among chromosomes turned out to produce an unexpected phenomenon in some cases. Most of the chromosomes in *Rhoeo*, *Oenothera*, and other species, assemble in rings that consist of nearly all the chromosomes of the complement. As early as 1917, O. Renner and later R.E. Cleland (1922, 1935) studied this situation that was found to be due to reciprocal translocations (A.H. Sturtevant and G.W. Beadle 1940).

Human chromosomes are not exceptions to these events. All known types of rearrangements also occur in the human complement. We have existed as a species for over 1 million years, and every human egg goes through thousands of cell divisions during the production of an adult, and yet human chromosomes have not disintegrated. Some of the rearrangements are a source of several diseases but others are not. One example is the translocation between chromosomes 9 and 22 that is involved in chronic myelogenous leukemia (J. Wang-Peng 1979). Another is the increase in the heterochromatic segment of the Y chromosome in different individuals (M.E. Drets and H.N. Seuanez 1974).

From the beginning geneticists tended to consider all the rearrangements, including crossing over, as random events. But as the molecular knowledge increased and a more exact analysis of the rearrangements became possible, order became evident. Soon the DNA sequences mainly responsible for this process were identified. These have been named transposons (S.N. Cohen and J.A. Shapiro 1980). They consist of a region of insertion flanked by duplicated sequences of which the bases are known. In maize retrotransposons have been sequenced and they are so common that they account for 50% of the total nuclear DNA (P. SanMiguel et al. 1996). Transposable elements have been identified in many species, including mammals, in which they were found to be nonrandomly distributed (H.A. Wichman et al. 1992). They were shown to occur in great amounts, and to have self-regulation systems that make them permanent components of the genome (E.G. Pasyukova et al. 1998, N. Borie et al. 2000).

STRUCTURAL CHANGE --1

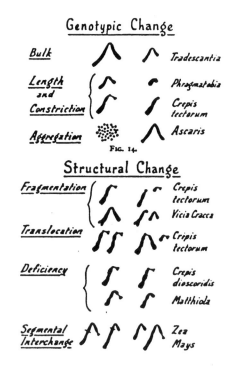

DARLINGTON 1937 - Diagram to show the kinds of change produced genotypically and structurally in the chromosomes.

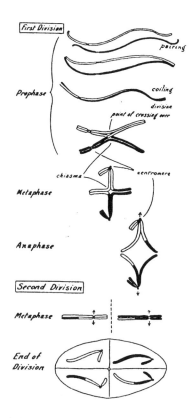

DARLINGTON 1932 - Diagram showing what happens to two mating chromosomes in the course of meiosis. Exchange of segments due to crossing-over.

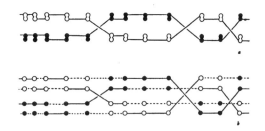

WHITE 1973 - BELLING's copy-choice hypothesis of crossing-over (1931). *a*, two-strand stage, in which the chromosomes (but not the chromonema) have replicated; *b*, four-strand stage. Maternal chromosomes white, paternal ones black.

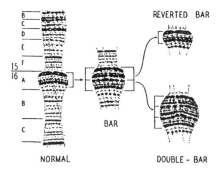

BRIDGES 1936 - The *bar* region of the X chromosome, from salivary glands of *Drosophila*, showing the duplicated section present in bar.

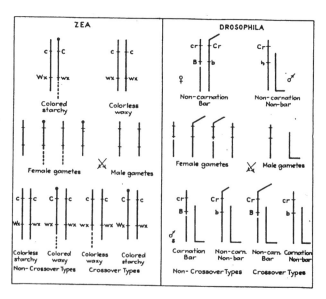

SHARP 1934 - Diagram illustrating the cytological demonstration of crossing over, based on the results of McCLINTOCK (1930) in maize and those of STERN (1931) in *Drosophila*.

STRUCTURAL CHANGE --2

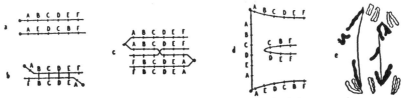

DOBZHANSKY 1941 - Crossing-over in inversion heterozygotes leading to dicentrics and acentrics, exemplified at anaphase of meiosis in *Lilium*.

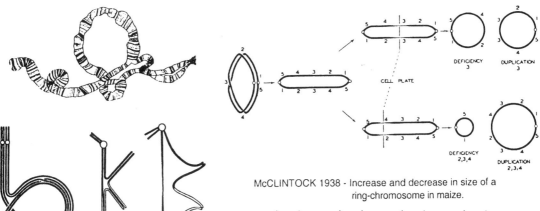

PÄTAU 1935 - Formation of a loop by pairing of inverted segments in *Drosophila* hybrids. Below, consequences of single crossing-over in such an inversion with the formation of a dicentric bridge and an acentric fragment.

McCLINTOCK 1938 - Increase and decrease in size of a ring-chromosome in maize.

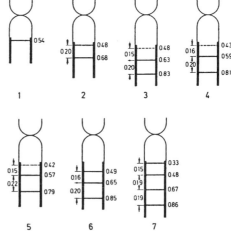

DRETS and SEUANEZ 1974 -
Diagrammatic representation of the relative size of the heterochromatic segment and relative band position along the long arm of the human Y chromosome.

WANG-PENG 1979 - Chromosomes 9 and 22 in human chronic myelogenous leukemia, showing the translocation of the distal segment of a 22(Ph^1) to the tip of long arm in a 9 chromosome.

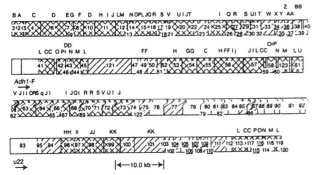

SANMIGUEL et al. 1996 - Composition and arrangement of repetitive DNA flanking the maize Adh1-F locus.

COHEN and SHAPIRO 1980 -
The duplication that attends the transposon insertion which itself causes a nine-nucleotide duplication.

Change of pattern

The change of chromosome pattern is such a dramatic event, that it is difficult to think, that one is looking at the same chromosome, when different cell types or cell stages are compared. The examples are as many as one would like to collect.

This phenomenon reveals itself in two different situations: One related to the division of the cell, the other resulting from tissue differentiation. A living organism has a phenotype as opposed to its genotype, but the chromosome also has its own phenotype which is separable from its genetic material.

In the first case, the chromosome gradually changes its phenotype, as it transforms itself during the cell division cycle. At interphase its structure is hardly recognized being invisible or nearly so. At prophase it is just a long and highly coiled thread and at metaphase becomes a compact cylindrical body (C.D. Darlington 1935, C.D. Darlington and K. Mather 1949). This change is most distinct at meiosis. The transformation that occurs from the elaborate chromomere pattern displayed at pachytene, to the compact body found at diplotene and metaphase I, is well known. This enormous process of compression is a general phenomenon in plant and animal species. It is equally evident in human chromosomes as displayed by the spiral structure of chromosome 2 at metaphase (Y. Ohnuki 1968).

The change of pattern in connection with tissue differentiation may be illustrated by the classic example described by C.B. Bridges (1935) in *Drosophila melanogaster*. He drew a gonial metaphase, in which chromosome number 4 is shown as a speck. At the side (and on the same scale) is the enormous chromosome number 4 in the salivary glands of this animal. This was a shocking surprise. No one could assume, that the chromosome had such an ability of transforming its pattern, yet maintaining its genetic information and its structural and functional organization. An interpretation of the process of polyteny, that partly explains this transformation, was given by H. Sass (1980). But much remains to be elucidated, in terms of the molecular processes that guide: 1) the packing of the DNA, 2) the folding of the proteins, 3) the differential formation of bands, 4) the compression and the decompression of the chromosome, as well as other events that still remain to be disclosed. A similar situation has been illustrated for the X chromosome of *Drosophila melanogaster* by C.D. Darlington and K. Mather (1949) in which the heterochromatic segment of this chromosome changes appreciably between the nerve ganglia cells and the polytene stage. This phenomenon was later studied by J. Gall (1973) who showed that the heterochromatin consists of satellite DNA.

The process of differential compression is also well illustrated in the nucleolar chromosome of tomato, which is shown changing its pattern from pachytene throughout meiosis and mitosis (S.W. Brown 1949).

One of the most dramatic changes in pattern that occurs in chromosomes takes place in spermatids. The chromatin becomes tightly packed as the chromosomes are pressed into the head of the sperm. This is seen in the cricket *Acheta* in the EM. The chromosomes form long "undulating curtains" that in transversal section appear as chromatin sheets of about 200 Å thickness (J.S. Kaye 1969).

In another respect the chromosome 4 of *Drosophila melanogaster* is also a paradigm. It is close to 0.2 of a micron at metaphase of mitosis and exceeds 15 microns in the salivary glands, with all its display of band pattern. For C.B. Bridges there was no hesitation in recognizing the two totally different structures as being the same chromosome. And why? Because, not only this chromosome, but all the others of the complement, were submitted to similar modifications. If these had not been simultaneous and coherent, chromosome 4, or the others, could not have been recognized with full accuracy.

CHANGE OF PATTERN.

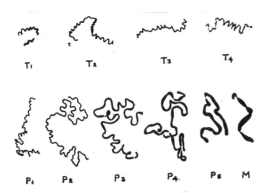

DARLINGTON 1935 - The uncoiling of relic spirals from telophase (T) to prophase (P) and metaphase (M) in *Fritillaria*.

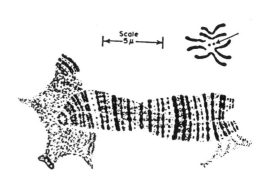

BRIDGES 1935 - The chromosome 4 of *D. melanogaster*, as it appears in the cells of the salivary gland and in gonial cells (upper right, and indicated by arrow).

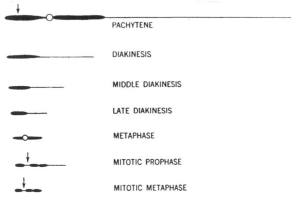

BROWN 1949 - Diagram to illustrate the changes in the chromatic (heavy portion) and achromatic (thin line) segments of the nucleolar chromosome in the tomato at different stages of meiosis and mitosis.

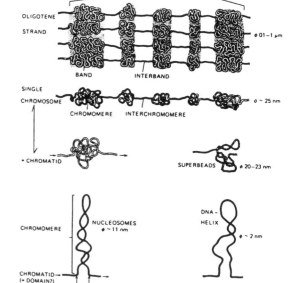

SASS 1980 - A schematic representation of the hierarchy of fibrillar organization levels in polytene, somatic and interphase chromosomes of *Chironomus* as seen in thin sections and spread preparations.

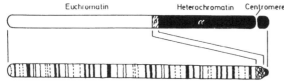

GALL 1973 - A comparison of a mitotic chromosome of *Drosophila* (above) with the corresponding giant polytene chromosome derived from it (below). The euchromatin of the mitotic chromosome gives rise to the regularly banded portion of the polytene chromosome. Most of the mitotic heterochromatin (α-heterochromatin) consists of highly repetitive satellite DNA which replicates little or not at all in the formation of the polytene chromosome.

KAYE 1969 - A transverse section through a spermatid nucleus of *Acheta*. Chromatin sheets about 200 Å thick occur in the peripheral zone.

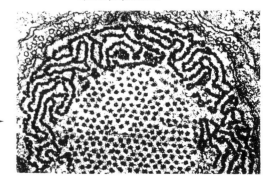

Change in size

By 1891 it was already noted that chromosomes varied in size from cell to cell within the same individual. This is a feature that cytologists seem to have taken for granted without having asked what kind of mechanism is involved in this transformation. Examples have accumulated, both in animals and plants, of this extreme variation in size, displayed simultaneously by all the chromosomes of the complement, or by some of them.

The chromosomes of the shark *Pristiurus* were drawn, to the same scale, at different stages of the development of the germinal vesicle showing quite different lengths (J. Rückert 1891). A similar example is found in the insect *Rhynchosciara angelae*. The entire nucleus of a somatic cell of this species was drawn at the same magnification as the tip of chromosome B at the puffed stage (C. Pavan 1958).

In the plant *Papaver rhoeas* (G. Hasitschka 1956) the same chromosomes are drawn in different tissues using the same magnification, and in *Phaseolus vulgaris* the giant suspensor chromosomes are compared with the minuscule ones found in root tips (W. Nagl 1974).

The size of chromosomes is also most different in closely related species of the genus *Crepis* (E.B. Babcock 1947), and in different species of the family Leguminosae (G.L. Stebbins 1971). In species of the family Amaryllidaceae the size decreases as polyploidy increases (B. Snoad 1956).

Within a complement the same phenomenon may assume extreme proportions but only in single chromosomes. In the beetles *Alagoasa* and *Walterianella* the sex chromosomes are gigantic in comparison to the autosomes despite the fact that we are observing a regular mitosis (N. Virkki 1964). Moreover, there is no indication of polyteny being involved at this stage in these gigantic chromosomes.

The variation in size observed under different situations is remarkable, but even more remarkable is that it is not an accidental process. It is repeated with accuracy once the molecular signals that lead to the formation of a given tissue are also repeated. The display, of an enormous or a minimal chromosome size, is unconditionally bound to the emergence of a given cell type, which obligatorily occurs during the organism's development and which in turn is determined by a series of well defined genes.

CHANGE IN SIZE.

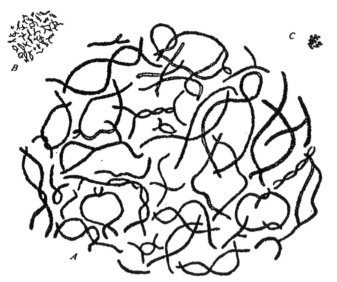

RÜCKERT 1891 - Chromosomes of the shark *Pristiurus* at different periods of egg development drawn to the same scale. A, period of maximal size (egg 3 mm); B, later period (egg 13 mm); C, at the close of ovarian life.

PAVAN 1958 - Entire nucleus from a young larva has a volume no bigger than the volume of a puffed band of the polytene chromosome. Camera lucida drawings of *R. angelae* at the same magnification. →

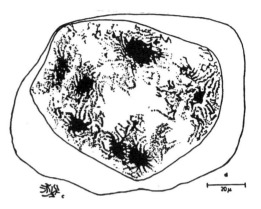

HASITSCHKA 1956 - *Papaver rhoeas* chromosomes at different stages. C, diploid metaphase; D, antipode cell with high degree of endopolyploidy.

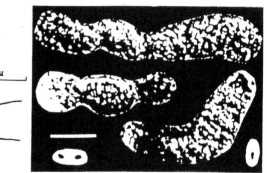

NAGL 1974 - The suspensor polytene chromosomes of *Phaseolus vulgaris* and three prophase chromosomes from the root tips. The bar indicates 10 microns.

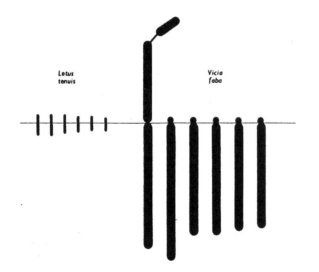

STEBBINS 1971 - Contrasting chromosome sizes in two species belonging to the same family (Leguminosae) and having the same basic number (n=6): *Lotus tenuis* and *Vicia faba*.

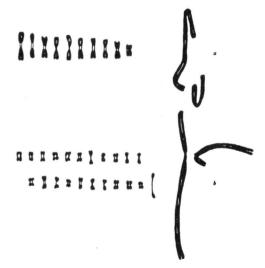

VIRKKI 1964 - Giant sex chromosomes in two species of beetles (a and b).

Change in number

Plant cytologists were the first to be confronted with a phenomenon, that turned out to be of general occurrence in plants, but rare in animals. As they started counting chromosome numbers, certain species happened to have complements consisting of 2, 3, 4, 5, 6 and even more than 10 times the chromosome number found in other organisms. Remarkable was that in most cases the increase was not just an accidental event, which occurred in one or two chromosomes, but it was the whole set that was duplicated as a unit. As cytologists found later, the duplication of the chromosome number could be experimentally induced by colchicine. This did not happen in few and discrete steps, but it occurred as a single event, in which all the chromosomes participated.

The genetic consequences of this event were even more surprising. A feature not often pointed out by cytogeneticists, was that the new plant or animal in which the chromosomes were duplicated, instead of being quite different continued to have essentially the same phenotype. A plant that became polyploid was slightly larger, and exhibited only minor deviations from its parents. A fly such as *Drosophila* that became polyploid continued to look like its ancestors with minor differences in phenotype. How was this possible in terms of gene expression? Suddenly the cell got all its genes in duplicated copies, and yet no serious disruption occurred in the final developmental result. At present it is known that, at least, some of the added genes are silenced, but we continue to be totally ignorant of the mechanisms that oblige the complement to behave in an ordered way, producing all the basic characters that characterize the species.

Other insects contain essentially the same genes as *Drosophila* and they are over 1 million species. Although triploidy and tetraploidy occur in *Drosophila* (E.W. Sinnott and L.C. Dunn 1939) polyploidy is a rare event in insects. The genetic dysfunction resulting from duplicated sex chromosomes has been considered a cause of the suppression of polyploidy in animals, but plant species also have sex chromosomes and in animals the autosomes also participate in sex determination (A.H. Sturtevant and G.W. Beadle 1940). The suppression is only partial since in humans triploid individuals (J.A. Book and B. Santesson 1960) and endoreduplication leading to the formation of diplochromosomes (C.B. Jacobson 1970) have been described.

In plants this genetic control seems to be less rigid. In many cases polyploidy results in a different species, but the new one still closely resembles the original parent. The problem may be even more complex. To start with, diploid organisms may not be "pure diploids", but may already contain several duplicated genes. The polyploidy would only magnify an originally established balance. In any case this only moves the question, concerning gene interaction, to one level below and as a consequence does not answer it.

G.D. Karpechenko (1928) was one of the first to study in detail the occurrence of polyploids in cultivated plants, such as *Brassica*, and to analyse the phenotypes resulting from its multiple chromosome sets and hybrids. M.B. Upcott (1936) described a polyploid series in the related species of the plant *Aesculus*. The polyploidy in plants may attain such extreme numbers as in the 630 bivalents shown at meiotic prophase (diakinesis) in the fern *Ophioglossum* (C.A. Ninan 1958).

The cell contains many alternative solutions. Instead of doubling the whole complement, in other situations it doubles only one of the chromosomes of the set. The result is what has been called a trisomy. This is an event that occurs in human cells, being the cause of severe pathologic conditions (E. Therman 1980). But it occurs in other animals and plants as well. A.F. Blakeslee and collaborators as early as 1937, were able to produce a series of 12 plants in *Datura*, each having one of the chromosomes of the complement in an extra dose. The fruits were slightly different but they could still not depart from the capsula shape. In *Taraxacum* the polyploid condition, as well as the suppression of meiosis leading to a different type of reproduction, were studied extensively by G. Gudjonsson (1946) and T. Sørensen and G. Gudjonsson (1946).

CHANGE IN NUMBER.

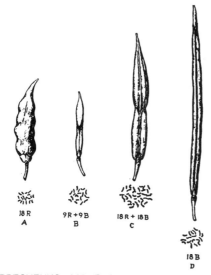

KARPECHENKO 1928 - Pods and somatic chromosomes of *Raphanus* (R) and *Brassica* (B) and their diploid and allotetraploid hybrids.

BLAKESLEE 1937 - The seed capsules of primary heteroploid mutants in *Datura*. The extra chromosome in each is shown diagrammatically.

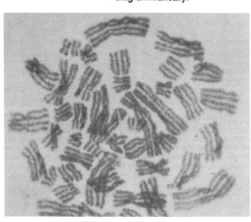

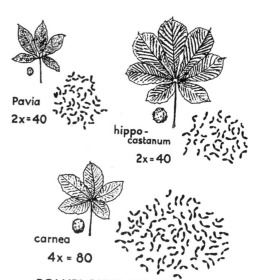

POLYPLOIDY IN ÆSCULUS

UPCOTT 1936 - The leaves, fruits and chromosomes of *Aesculus carnea*, which is a tetraploid, and its diploid parents.

NINAN 1958 - Meiotic prophase (diakinesis) in a sporocyte of *Ophioglossum reticulatum*, showing about 630 bivalents.

STURTEVANT and BEADLE 1940 - Triploid intersex of *Drosophila melanogaster*. At right, metaphase chromosomes (after Bridges).

JACOBSON 1970 - Human diplochromosomes produced by replication without separation of the new sister chromatids.

Change in function

The chromosome plasticity takes other forms. One of them is the change in function. Two major events are involved in this process: (1) the variation in the position and number of the DNA sequences along the chromosome; and (2) the way the chromosome message is modified after it has been released from the DNA. Both situations result in a functional message different from that written in the original chromosome sequence.

Position effects were extensively studied during the 1930s and until the 1950s. They furnished the first cases of the change in gene function due to the alteration of a gene's position in the chromosome or as a consequence of the transfer of other DNA sequences to the vicinity of another gene. This area of research then became out of fashion.

Until 1967 most structural genes and other DNA sequences were considered to occur as single copies, or in exceptional cases, in a few discrete repeated numbers. It was the work of R.J. Britten and collaborators (R.J. Britten and D.E. Kohne 1970) that disclosed in 1967–1969 that chromosomes from protozoa to humans contained repeated DNA sequences. P.M.B. Walker et al. soon found (1969) that in rodents, such as the mouse, there was a class of DNA sequences which occupied 10 to 15% of the total genome that consisted of 10^6 reiterations. In their original paper they had already pointed out that this sequence increase should modify chromosome function. At present redundancy of DNA sequences is known to be a property of every genome. The repeated sequences have been divided into several categories. (1) Simple sequence families consisting of 1,000 to 10 million copies (each unit being made up of a few to a hundred nucleotide pairs), examples are DNA satellites. (2) Multiplicational families encompassing up to 10,000 copies of a gene (approximate length 100 to 1,000 nucleotide pairs), examples are ribosomal RNA and histone genes. (3) Informational families containing only a few to some hundred individual genes with varying length, examples are the genes for many enzymes as well as for haemoglobin.

The generation of multigene families has been attributed to the occurrence of ubiquitous interspersed DNA segments such as the so-called *Alu* sequences. These are quite common in the human genome and consist of about 300 base pairs, in perhaps 500,000 copies, accounting for as much as 5% of human DNA. The *Alu* sequences are readily transposable and their name derives from their DNA being cleaved by the restriction endonuclease *Alu1*.

More important, however, are the magnification and amplification mechanisms which sense and adjust the number of genes. These are keys to our understanding of the organization of the chromosome as a unit.

F.M. Ritossa and coworkers discovered a phenomenon in *D. melanogaster* which they called magnification (F.M. Ritossa et al. 1971). A wild-type locus consisted of 130–300 ribosomal genes. If the genome contained less than 130 genes for ribosomal RNA, its synthesis was inadequate, and the mutated bobbed phenotype developed. The magnification appeared in the progeny of strongly bobbed males and involved a rapid accumulation of ribosomal RNA genes unstably bound to the chromosome. By this process the amount of ribosomal RNA genes was rapidly adjusted to the normal number (F.M. Ritossa 1973). Hence, magnification consisted of a sudden increase in the initial ribosomal gene redundancy pertaining to a deleted locus. The increase was inheritable and caused in one or a few generations the achievement of the wild-type condition or of a gene number still higher than the wild type. Later K.D. Tartof (1971, 1974) studied wild-type *D. melanogaster* females which had about 250 ribosomal RNA genes in each nucleolus organizer region of the two X chromosomes. When this same region was present in a single dose the number of ribosomal RNA genes increased to about 400. This occurred during the ontogeny of a single generation and it happened in these females and in X/0 males. K.D. Tartof concluded that the phenomenon of magnification was in several respects analogous to amplification and that it represented a mechanism capable of "sensing" a deficiency in the number of ribosomal RNA genes and of introducing the necessary adjustment. This sensing mechanism remains today as elusive, in molecular terms, as it was at that time. We are just starting to discover that proteins may function as messengers along the chromosome, furnishing the necessary information to alter or to maintain a given function. These proteins, function as switches, that can turn on a series of genes affecting the body pattern (M. Mlodzik and W.J. Gehring 1987).

Mini- and microsatellites are another type of repetitive DNA. As T.A. Brown (1999) points out "their function is mysterious". They reveal a unique genetic profile that can be established for every person and that is used as a tool in forensic medicine.

Gene amplification denotes a process by which specific genes are synthesized preferentially in relation to others, leading to a quantitative increase of one or some particular gene sequences. The amplified genes may or may not remain associated with the chromosomes. The main difference between reiterated (or redundant) DNA and amplification, is that the latter represents an increase in gene number which is confined to specific tissues, and which does not always involve integration into the chromosome.

The occurrence of amplification in oocytes has a long cytological history going back to the turn of the century (A. Giardina 1901). It was, however, in the 1940s and 1950s that the evidence started to accumulate at the cytochemical level. This led to its demonstration by molecular methods. Direct evidence for the amplification of ribosomal RNA genes in early amphibian oogenesis was provided by *in situ* hybridization and gene isolation. The haploid chromosome set of a somatic cell of *Xenopus laevis* contained 450–800 copies each of 18S and 28S ribosomal genes (H. Wallace and M.L. Birnstiel 1966). Oocytes at the meiotic prophase contained 1,000–4,000 times more DNA than the somatic cells of the same individual. This was a massive form of gene amplification (D.D. Brown and I.B. Dawid 1968). Amplification was demonstrated for other genes as well as in other animals. Besides amphibians it was extensively studied in the cricket *Acheta domesticus* by means of *in situ* hybridization, gene isolation, DNA-RNA hybridization and measurements of the amount of DNA in chromomeres. The extra copies of 18S and 28S ribosomal RNA genes formed a large DNA body in the oocytes but not in spermatocytes. The amplification was tissue specific as in *Xenopus*. After they were used in RNA synthesis the extra copies left the chromosome in an organized way. The number of nucleotide pairs released from the chromomeres has been measured (A. Lima-de-Faria et al. 1968, 1973).

Of special relevance to the sensing and adjusting capacity of the amplification process are the results on a mutant in *Xenopus* which in the homozygous form has no nucleolus organizers on either homologous chromosome. The embryos of this anucleolate or nucleolus-less mutant do not synthesize any ribosomal RNA. *Xenopus* oocytes from females heterozygous for the mutant nucleolus-less contain the same number of nucleoli, and the same amount of extra ribosomal DNA, as wild-type oocytes (E. Perkowska et al. 1967). Somatic cells, however, have lost half of their ribosomal RNA genes and do not compensate for the missing ribosomal DNA complement (M.L. Birnstiel et al. 1966). H. Tobler (1975) concludes that a control mechanism must exist in the oocyte which is able to sense the number of ribosomal RNA genes and to adjust its number of replications accordingly. Although magnification was analysed in an insect (*Drosophila*) and amplification mainly in amphibians and in another insect species, such as *Acheta*, the genetic mechanism appears to be basically the same. The cases studied so far have disclosed an ability of the chromosome to recognize the number of its specific genes and to adjust its gene content in a way that establishes the number that the chromosome complement finds appropriate.

The designation "Fragile sites" has been given to points at which chromosomes break non-randomly. The name derives from the fragile X syndrome that is associated with human mental retardation. They are actually dynamic mutation sites and have been considered to have a role in the onset of cancer (J.L. Mandel 1997).

Typical eukaryotic protein-coding genes consist of exons, introns and flanking regions (W.-H. Li and D. Graur 1991). Pseudogenes are the "false doors" that the genome keeps hidden in its many compartments. They are genes bearing a close resemblance to other genes (in terms of nucleotide sequences) but usually not functional. Normal transcription or translation has been mainly prevented by duplications or deletions. An example is the globin genes which appear to have originated by duplication and were silenced by point mutations and other rearrangements. However, they have been maintained in the genome for millions of years. Another class is formed by the processed pseudogenes, that lack introns, and are flanked by direct repeats of 10 to 20 nucleotides. In humans there are 20 pseudogenes that are supposed to be derived from actin and beta tubulin messenger RNAs. Retrotransposition is considered to be involved in the emergence of pseudogenes because these are flanked by direct repeats (C.D. Wilde et al. 1982).

What is significant in connection with genome plasticity, is that "camouflaged" sequences of ribosomal genes, telomeres and centromeres, which are present as additional copies, but which are silent, suddenly become active when rearrangements locate them in new positions (as mentioned in previous chapters). It is also well known that genes encoding new proteins can be created by the recombination of exons or by the fusion of duplicated genes (N. Maeda and O. Smithies 1986).

The functional plasticity of the genome becomes even more striking when one analyses the final messages obtained from the chromosomal DNA. Splicing is the process whereby noncoding sequences, or introns, are removed from precursor RNA molecules (H. Lodish et al. 1995). This can occur as self-splicing or as the splicing of other molecules. Self-splicing led to the discovery of RNA enzymes in 1981 by K. Kruger et al. But of special interest is that alternative splicing can occur in different tissues. Different RNAs can be transcribed from the same gene, as in the case of the pyruvate kinase M gene, giving rise to different proteins (H. Muirhead et al. 1986, L.A. Fothergill-Gilmore 1991).

This is a virgin area where so much remains to be discovered. The fact that RNA can behave as a protein adds a new functional dimension to this molecule. RNA is the most versatile of the cell's genetic molecules. It can be: (1) the genetic material, as in RNA viruses, (2) the messenger of the code, as in messenger RNA, (3) the director of protein synthesis, as in ribosomal RNA, (4) the key molecule in amino acid assembly, as in transfer RNA and (5) a protein with catalytic properties.

Among the surprises was the finding in 1990, that proteins have the ability to excise themselves building a smaller protein out of a large precursor. This is a self-catalytic process analogous to the trans-splicing of introns in RNA (N. Liu et al. 1998).

Genome imprinting is another phenomenon that alters gene function. It has been defined as the transcription of only one allele at a locus, dependent on the parental origin of the allele. This event affects both the expression and the transmission, of particular genes, leading to a violation of Mendel's laws (F.P. Villena et al. 2000). In humans this differential expression of paternally and maternally derived genes is affected by long-range mechanisms (R. Allshire and W. Bickmore 2000).

CHANGE IN FUNCTION --1

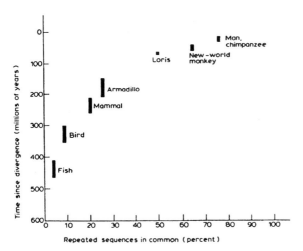

BRITTEN and KOHNE 1970 -
Relationship between the similarity of repeated-DNA sequences and evolutionary history.

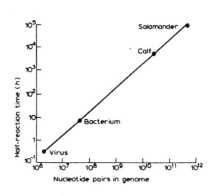

BRITTEN and KOHNE 1970 -
Time for reassociation of single copy DNA varies with the size of the genome and therefore increases with the complexity of the organism.

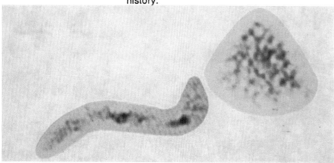

LIMA-DE-FARIA et al. 1973 -
Upper Figure. Amplification of ribosomal RNA genes in chromosome 6 of the cricket *Acheta* in oocytes at pachytene. The amplified DNA body (right) contains 2.5×10^9 nucleotide pairs. Lower Figure. Volumetric display of DNA amount in chromosome 6 and the DNA body.

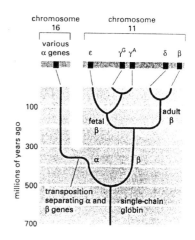

MAEDA and SMITHIES 1986 -
Evolution of the globin gene family. A gene duplication of the gamma-chain gene produced gamma G and gamma A, which are fetal beta-like chains of identical function. The globin genes are located in the human chromosomes 16 and 11.

FOTHERGILL-GILMORE 1991 -
Alternative splicing in different tissues of RNA transcribed from the pyruvate kinase M gene. The gene has 12 exons. Exon 9 is specific to the muscle isoenzyme and 10 to the kidney isoenzyme. ⟶

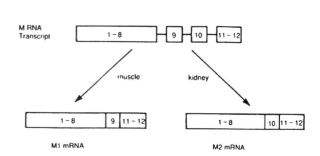

CHANGE IN FUNCTION --2

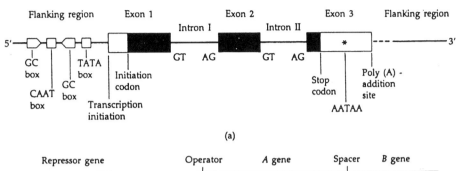

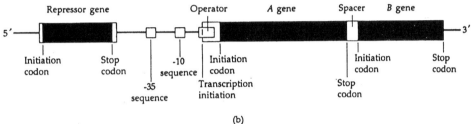

LI and GRAUR 1991 -
(a) Schematic structure of a typical eukaryotic protein-coding gene. (b) Schematic structure of an induced prokaryotic operon.

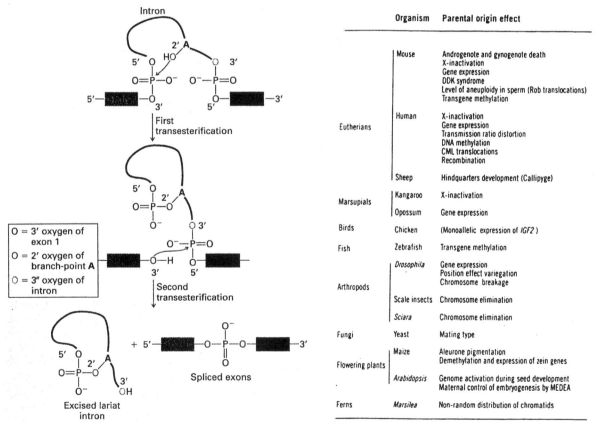

LODISH et al. 1995 -
Splicing of exons in pre-messenger RNA occurring via two transesterification reactions. The intron is released as a lariat structure and the two exons are joined.

VILLENA et al. 2000 -
A list of parental origin effects to illustrate the diversity of organisms in which they are observed.

PART VIII

CHROMOSOME MODELS AND WHAT THEY DO NOT TELL US

The models

Models have advantages and disadvantages.

The positive side of a diagram is that it simplifies and concretizes a dispersed body of knowledge that had not before been condensed into a single picture. Another advantage is that by formalizing several concepts it obliges researchers to find out whether the qualifications assigned to the model are actually correct. In this way it promotes further research which is targeted at validating or invalidating the model.

The negative aspect is that a diagram is always a simplification. Moreover, it often includes acquired misconceptions or factual errors which tend to be perpetuated in this way. This makes it dangerous because, time and again, it is reproduced, being accepted as the final interpretation of a given structure or phenomenon.

This is exactly the case with chromosome models. They have been valuable because they led to further research directed at testing their validity, and they had a negative effect because they perpetuated misconceptions.

Although chromosomes had been discovered several years before the initiation of the 20th century it was not until 1928 that the first model appeared. This date is not accidental. Nucleic acids had been known to be part of cells since 1869 as a result of F. Miescher's chemical work, and proteins had been identified in the time of J.J. Berzelius (1838). However, it was only in 1923 that R. Feulgen and H. Rossenbeck demonstrated by means of a chemical reaction that chromosomes contained this acid. It was also at this time that proteins started to be understood as giant molecules when compared with other cell chemicals. As genetics advanced, the central idea that developed, was that nucleic acids were too simple molecules to harbour the complexity of the gene. Proteins ought to be the molecules responsible for the transmission of hereditary characters since they were large and complex. This idea was perpetuated until 1953, when the J.D. Watson and F.H.C. Crick model of DNA was proposed. This means that it was widely accepted for 30 years.

The experiments of O.T. Avery and collaborators (1944) demonstrating that, in bacteria, DNA was the genetic material, were considered to be of no relevance for higher organisms. At that time it was not known whether bacteria had chromosomes, or whether they had genetic mechanisms comparable to those of plants and animals. Moreover, as early as 1945 W.T. Astbury had proposed a model of DNA in which the bases were stacked on top of each other, their distance being 3.4 Å, but his work as well had no immediate impact.

N.K. Koltzoff (1928) in the Soviet Union was a visionary. He synthesized the little knowledge available at the time into a diagram with several novel features. The model included the two chromatids of a chromosome as basic units of replication, a feature that had not been clearly defined before. Also, he represented the chromosome as consisting of a large number of protein molecules which were part of its body and of its cellular environment, emphasizing in this way its molecular activity. The most original feature was that he represented each chromatid as consisting of paralleled aligned molecules which built two identical fibers, a prerequisite for chemical copying.

It is striking to note that most chromosome models that appeared in the thirties, fourties, and even in the fifties, lacked the molecular component and did not include the chromosome replication. This is the case with the diagrams of chromosomes published by E. Heitz (1935), L.W. Sharp (1943), F. Schrader (1944), E.D.P. De Robertis et al. (1948), A. Engstrom and J.B. Finean (1958).

What is new in these models is that they include the spiral structure of the chromosome which had been so well depicted and photographed by many different cytologists, both in plants and animals. Also by this time, three regions of the chromosome had been found to be of general occurrence. These were the centromere (M. Navashin 1934), the nucleolus organizer (E. Heitz 1931) and the telomeres (H.J. Muller 1940).

These models also taught us how entrenched ideas, that were not well established, took time to die out. E. Heitz's model of 1935 includes the matrix as a main component of the chromosome body. This structure is included as late as 1958, but its existence was never established, being abandoned in the 1960s.

The chromosome pellicle was considered a necessary structure that enveloped the chromosome body delimiting it well from the nuclear sap and from the cytoplasm. The pellicle is included in the models of F. Schrader (1944) and E.D.P. De Robertis et al. (1948). This thin sheet was actually never observed and soon it became evident that the chromosome had no limiting membrane.

The number of chromatids present in a chromosome at metaphase was for years, a source of controversy, but this problem faded away when endoreduplication was discovered.

One of the first models to include the different levels of organization from the gene to the metaphase chromosome was that of C.D. Darlington and K. Mather (1949). The nucleotides (PSB, phosphate-sugar-base) were included in the scheme for the first time, but the gene (G) was located not as a part of their chemical composition, but was drawn at the side, anchored on the protein fiber (F). Hence the gene continued to consist of protein. But this model represented a big step forward, for it not only described all the organization levels, but it included the dimensions of the various chemical and structural components.

In the 1960s and 1970s other problems arose. The introduction of tritiated thymidine, as a tool in the study of chromosome and DNA replication, raised new questions. The use of the electron microscope, which allowed the visualization of single DNA and RNA molecules, also created novel ideas. As a result the internal organization of the chromosome at the molecular level became the focus of much attention.

But an uncertain feature was the replication of the DNA molecule along the chromosome. It was difficult to conceive that such a molecule, could be so long, as to be continuous throughout the whole chromosome length. Could it replicate as a unit? One was led to consider that DNA may not be continuous throughout the whole chromosome but small DNA molecules were linked by proteins that functioned as bridges, keeping the whole as a unit (A.E. Mirsky and H. Ris 1947, 1950, H. Bush 1962). Even after the use of tritiated thymidine in the study of DNA, replication models were devised, in which protein linkers were included between DNA molecules. But by the middle 1970s this concept was dead. T.D. Petes (1974), and others, had measured the length of the DNA fiber in single chromosomes and demonstrated the continuity of the DNA molecule along the whole chromosome body (R. Kavenoff and B.H. Zimm 1973). At about the same time the DNA of higher organisms was found to consist of well defined replication units with distinct initiation sites (H.J. Edenberg and J.A. Huberman 1975).

It also became evident that the DNA molecule crossed the centromere from one arm to the other, but the division cycle of this region remained a source of confusion. Cytologists, who obtained good preparations of chromosomes, could see that the centromere, which was attached to the spindle, was divided at metaphase and that it was the proximal regions of the arms (usually heterochromatic) that held the chromatids together, being the last regions to divide at anaphase. This is shown in the models of L.W. Sharp (1943) and of E.D.P. De Robertis et al. (1948) and was a feature later found to be of general occurrence in plants and animals, being included in the model of D.E. Comings (1972). However, many recent textbooks still present models with the centromere undivided at metaphase, an archaic notion based on the defective cytology of the 1930s.

Advancements in the study of the tertiary and quaternary structure of proteins led to the discovery of the packing of the DNA with histones resulting in the building of nucleosomes. The result was a better representation of the successive orders of packing, taking place between DNA and the coiled metaphase chromosome (D.E. Comings 1977, B. Alberts et al. 1994).

A structural feature that had been discarded since the 1940s was the chromosome scaffold which consisted of nonhistone proteins. Since H. Ris and A.E. Mirsky's classic experiments (1947, 1950), the DNA and the histones could be removed from a metaphase chromosome with the unexpected result that it continued to maintain its basic shape. These proteins had been forgotten and were not included in any models. It was the electron microscopy work of J.R. Paulson and U.K. Laemmli (1977) that showed that the scaffolding proteins were the molecular component that maintained the chromosome's framework. As a result the scaffold proteins became incorporated in the new models (A.J.F. Griffiths 1993). The DNA and the nucleosomes were attached to this nonhistone protein (having also its own spiral structure).

Recently the models of chromosomes have taken on a rather different form due to the base sequencing of entire genomes in plants, animals and humans (2000, 2001). Along the chromosome body are now represented the frequencies of different types of DNA sequences such as: tandem repeats, centromere associated repeats, telomere sequences, dispersed TY1/copia, LINEs (long interspersed elements), SINEs (short interspersed elements) and genes (T. Schmidt and J.S. Heslop-Harison 1998).

CHROMOSOME MODELS -- 1

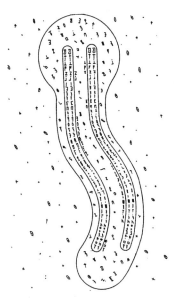

KOLTZOFF 1928 -

The first model of a chromosome depicting its chemical composition as well as its separation into two chromatids as a consequence of a biochemical process. Each chromatid was considered to be composed of aligned molecules building two identical fibers. In 1928 the genetic material was supposed to consist of proteins. Nucleic acids were thought to be too simple to harbour the complex structure of the gene.

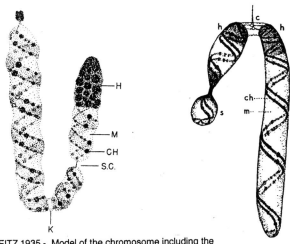

HEITZ 1935 - Model of the chromosome including the main cytological features of its structure. CH, chromomere; H, heterochromatin; K, kinetochore or centromere; SC, secondary constriction (sometimes identical to nucleolus organizer); M, matrix. Until the 1960s the existence of a matrix was under debate.

SHARP 1943 -

Diagrammatic representation of a chromosome at anaphase of mitosis. C, centromere; CH, chromonema; H, heterochromatin; M, matrix; S, satellite with adjacent nucleolus organizer (constriction). The matrix was considered to be a substance found between the chromonema.

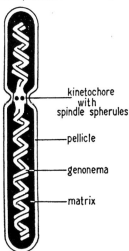

SCHRADER 1944 -

Model of the chromosome introducing several new features. At that time the number of chromonemata in a chromosome was under dispute (two being shown). It was also not known whether the coiling of the chromonemata was independent or relational. Attention is called to the presence of spindle spherules in the kinetochore. Besides the matrix, the chromosome is shown to have a pellicle, that enveloped the whole body. It turned out later that the pellicle does not exist.

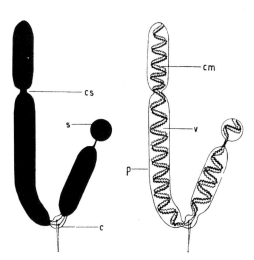

DE ROBERTIS, NOWINSKI and SAEZ 1948 -

The morphology and internal structure of a chromosome. Left, morphology as observed at anaphase. Right, internal structure of the same chromosome following treatment. CS, secondary constriction; C, centromere or primary constriction; S, satellite; CM, chromonema showing the major and minor spirals; P, pellicle; V, sheath or matrix. The centromere is shown divided.

CHROMOSOME MODELS -- 2

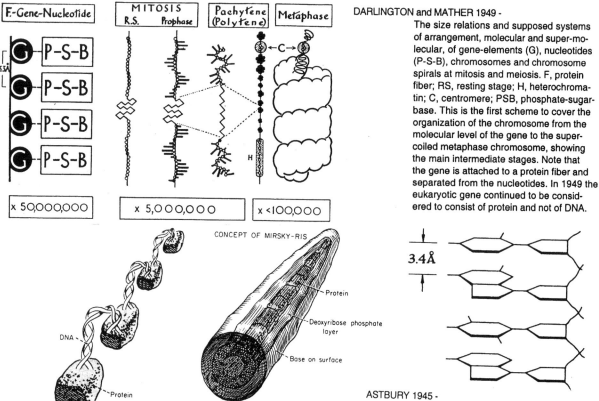

DARLINGTON and MATHER 1949 -
The size relations and supposed systems of arrangement, molecular and super-molecular, of gene-elements (G), nucleotides (P-S-B), chromosomes and chromosome spirals at mitosis and meiosis. F, protein fiber; RS, resting stage; H, heterochromatin; C, centromere; PSB, phosphate-sugar-base. This is the first scheme to cover the organization of the chromosome from the molecular level of the gene to the super-coiled metaphase chromosome, showing the main intermediate stages. Note that the gene is attached to a protein fiber and separated from the nucleotides. In 1949 the eukaryotic gene continued to be considered to consist of protein and not of DNA.

MIRSKY and RIS 1947 -
Model of a chromosome in which protein blocks serve to bind ends of DNA helices together. The discontinuity of the DNA molecule along the chromosome was a prerequisite to explain its DNA replication. The replication of a single DNA molecule covering the whole chromosome was considered an impossibility.

MIRSKY and RIS 1947 and 1950 -
Molecular organization of the segment of a chromosome. The protein is the inner core and the DNA is the surface structure with its bases on the surface of the chromosome body (From BUSCH 1962).

ASTBURY 1945 -
The base-stacking structure of DNA. In this structure the deoxyribose ring is coplanar with its attached purine or pyrimidine ring, both sugar and base being perpendicular to the long axis of the polynucleotide chain molecule.

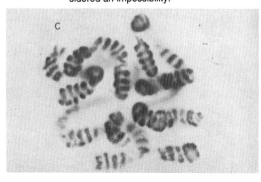

DARLINGTON 1950 -
Left, light micrograph of meiotic metaphase chromosomes in *Fritillaria* showing the spiral structure. Right, drawing of a bivalent chromosome at the same stage and in the same species exhibiting the major and minor spirals.

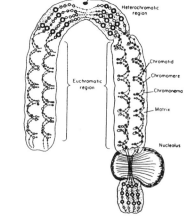

ENGSTRÖM and FINEAN 1958 -
Schematic representation of a chromosome. This model is not based on a direct observation of the chromosome and as a result contains several errors. The centromere is represented as a single dot and the nucleolus is included in an apparently anaphase chromosome (a stage at which the nucleolus has disappeared).

CHROMOSOME MODELS -- 3

COMINGS 1972 -
    Single-stranded model of chromosome structure. A single DNA-protein fiber is considered to begin at one telomere, to fold upon itself, to build up the width of the chromatid and eventually to progress to the opposite telomere without lengthy longitudinal fibers, with no central core and no half- or quarter-chromatids. The centromere is represented well divided in the metaphase chromosome.

Species	Strain	Idiogram	Largest chromosome to scale	Molecular weight of largest DNA
melanogaster	wild type		—	41×10^9
	inversion		—	42×10^9
	translocation		—	58×10^9
hydei	wild type		—	40×10^9
	deletion		—	24×10^9
virilis			—	47×10^9
americana			—	79×10^9

KAVENOFF and ZIMM 1973 -
    The DNA molecule runs the entire length of a chromosome. This conclusion was supported by the finding that the cytological chromosome size agrees with the molecular weight of the largest DNA molecule for *Drosophila*.

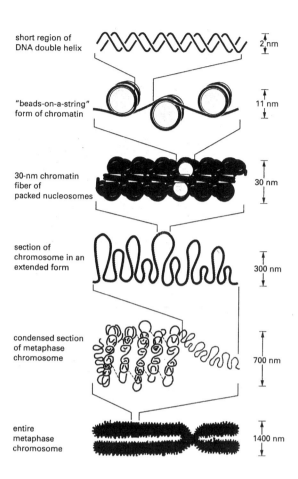

ALBERTS et al. 1994 -
    Model of chromatin packing showing some of the successive orders of organization postulated to give rise to the highly condensed mitotic chromosome. The DNA becomes packed with histones in the form of nucleosomes which associate building successive levels of spirals. In this model nonhistone (or scaffolding) proteins are not represented.

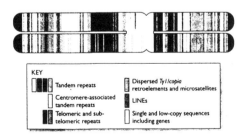

SCHMIDT and HESLOP-HARISON 1998 -
    Model for the positioning of different types of repetitive DNA in a typical plant chromosome. The positions of the dispersed Ty1 copia retroelements and microsatellites are shown in the lower pair of chromatids only. The original is in colour.

What the models do not tell us

But let us now try to see what the available models lack. Surprisingly there is an impressive amount of information on chromosome properties which has been left out of these graphic representations. This is apparently due to the absence of a coherent picture that integrates the interactions between the different regions. The information that is missing mainly has to do with the function of the chromosome as a whole. Without it, it seems difficult, if not impossible, to discover the mechanism responsible for the maintenance of chromosome integrity.

Polarity in the gene and the chromosome
The polarity already present in the DNA molecule (J.D. Watson and F.H.C. Crick 1953) is found in other forms at higher levels of chromosome organization. Gene polarity was disclosed by mutant analysis of the operon (F. Jacob and J. Monod 1961, N.C. Franklin and S.E. Luria 1961). In an operon the messenger RNA was transcribed from the operator to the distal end, and polar mutations resulted in loss of enzymatic activity. But polarity did not occur solely in the operons of bacteria, it was also found in eukaryotic chromosomes where it became evident in DNA replication (J.A. Huberman and A.D. Riggs 1968), RNA transcription (K. Gross et al. 1976) and DNA recombination (H.L.K. Whitehouse 1972). Polarity may cover a few nucleotides but may extend throughout a series of loops in the lampbrush chromosomes and even throughout the entire chromosome at meiosis (H.G. Callan 1963).

C.D. Darlington (1937) was the first to point out that many chromosome properties behaved as if they would have their center of origin at the centromere. It is this property that led him to coin this term, *i.e.* the central part of the chromosome. The terminalization of chiasmata is an example of such a phenomenon (M.J.D. White 1973).

Position effects extend throughout whole arms
The interaction between chromosome regions, situated in different parts of the same chromosome, was one of the main areas of study from the 1930s to the 1950s extending in some cases into the 1970s. Position effects were analysed mainly in *Drosophila* but were found as well in plants, mice, humans and other organisms. In *Drosophila* the interaction between genes spread over 50 bands in the salivary gland chromosomes (M. Demerec 1941). This meant that the suppression effect on gene activity extended for 15×10^5 nucleotide pairs (G.T. Rudkin 1972). But other DNA sequences had an action that covered over 65 bands (I.J. Hartmann-Goldstein 1967). In some cases the gene effect extended over 100 bands covering the whole chromosome arm (M.M. Green 1962).

The study of these interactions at the molecular level disclosed that the nucleic acid content of chromosome bands was dependent on their position within the chromosome (T.O. Caspersson 1950) and the DNA replication of genes could be changed by DNA sequences situated over one million nucleotides away (E.V. Ananiev and V.A. Gvozdev 1974).

These interactions, affecting gene expression and molecular function, are crucial for the understanding of the organization of the chromosome as a whole.

Clustering of genes with similar functions and the occurrence of gene territory
The occurrence of gene clustering first became evident in bacteria. M. Nomura (1977) could show that the genes for ribosomal proteins were located at the side of each other. This phenomenon was also found in eukaryotes such as *Neurospora* (M.E. Case and N.H. Giles 1971), flowering plants (C.M. Rick 1971), the mouse (M.T. Davisson and T.H. Roderick 1980) and in humans (V.A. McKusick 1971). The problem of gene clustering is directly related to that of gene territory. Already E. Heitz (1932), in his work on the distribution of heterochromatin, pointed out that it tended to occur at sites related to the position of the centromere and telomeres. This type of territory turned out to be of general occurrence. Specific genes, such as the genes for 28S and 18S ribosomal RNA were found to occur in 86.6% of the cases near telomeres, in distantly related organisms such as algae, flowering plants, animals and humans. Other DNA sequences also tended to have specific territories (A. Lima-de-Faria 1979).

The chromosome maps presented by the Human Genome Project (G.D. Schuler et al. 1996 and J.C. Venter et al. 2001), revealed that genes are not distributed evenly throughout the genome, each chromosome having gene-dense and gene-light regions and some chromosomes having gene clusters concentrated to specific regions.

All this preliminary information indicates that much must be learned about interactions occurring along the centromere-telomere region, building what has been called the chromosome field (A. Lima-de-Faria 1980, H. Scherthan et al. 1987).

One wonders what chromosome models will look like 30 years from now. We will probably have reached the electronic level of the bases and of the amino acids. It is expected that the models will have much larger dimensions to accommodate the proton-electron interactions that direct protein and DNA evolution. Moreover, these models may represent the chromosome in a novel way as a fully integrated body, where most DNA sequences interact with each other in a concerted form collaborating in the preservation of its identity.

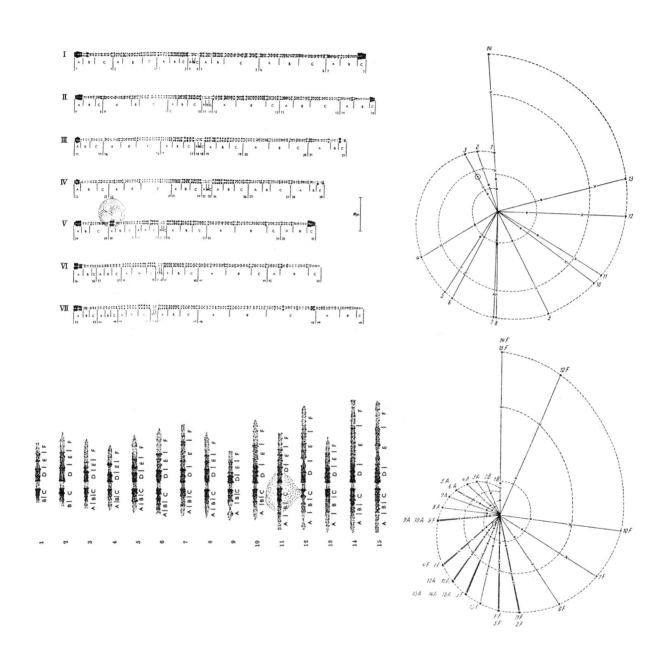

LIMA-DE-FARIA 1952, 1954, 1983 -
A chromosome field resulting from the interaction between centromeres and telomeres was proposed in 1954 based on the ordered distribution of the chromomere pattern found in plant species. This pattern has the following characteristics in 73 species investigated: (1) It starts on both sides of the centromere. (2) The chromomeres decrease in size along each arm towards the telomere (an arm being divided into three regions). (3) The decrease in size is dependent on the centromere-telomere distance (and is affected as well by the location of the nucleolus, as in the short arm of chromosome 5 of rye). (4) The distribution of all arms and arm regions builds a spiral, both in *Secale cereale* (rye n = 7, upper figure) and *Agapanthus umbellatus* (n = 15, lower figure). As is known from other biological phenomena the formation of a spiral is indication of a coherent and dynamic process which in this case is displayed by the centromere-telomere sequence. (5) The phenomenon is of general occurrence since it occurs in unrelated plant families and is independent of the cell stage (pachytene in rye, prophase II of meiosis in *Agapanthus*). Fifty years ago there was no immediate explanation for this phenomenon but at present studies, emerging from different laboratories, are revealing molecular interactions occurring within the telomere-centromere segment (H.A. Wichman et al. 1991, S. Henikoff 1992, P. Slijepcevic 1998, D. De Bruin et al. 2000) (see pages 34, 75, 134).

PART IX

EPILOGUE

Where did the chromosome come from?

The origin of the cell and of the chromosome are not known
Much research has been carried out concerning the origin of life with its implications for the origin of the chromosome. Although some of the basic processes of protein and nucleic acid assembly have been elucidated, the exact molecular processes involved in the emergence of the cell and especially of the chromosome remain elusive (A.I. Oparin 1961, S.W. Fox 1965, M.G. Rutten 1971, L.E. Orgel 1973, L. Margulis 1981, A.B. Martin and P.G. Schultz 1999, P. Schimmel and S.O. Kelley 2000). We simply do not know what atomic interactions led to the self-assembly processes which put together the structural proteins, the histones, the RNAs and the DNAs into a coherently organized organelle which is called the chromosome.

Nucleosomes, the basic constituents of chromosomes, are formed by the self-assembly of DNA with the different histones (P. Oudet et al. 1975) and the interactions between these macromolecules have been established at the atomic level (K. Luger et al. 1997) but much remains to be learned concerning the atomic processes which shaped the chromosome as a functional unit.

The unique position in the Periodic Table of the atoms used in the construction of the cell and the chromosome
The question that then arises is what kind of atoms were initially employed in the construction of the chromosome. This question can be best answered by considering the properties of the known chemical elements. These are 106, 102 of which are well characterized, starting with the simplest, which is hydrogen, and finishing with the most complex, nobelium (N.N. Greenwood and A. Earnshaw 1989).

Surprisingly there was a preference in atomic combination when the cell and the chromosome were built. As early as 1945, A. Frey-Wyssling noted that the chemical elements obligatory for cell construction and function were only 25, having a regular distribution in the Periodic Table. All were light elements with an atomic number of less than 35, building mainly a line from carbon to argon. But when it came to the chromosome the restriction became still more evident.

The constituents of chromosomes, *i.e.* the nucleic acids and proteins, consist of atoms which occupy a particular location in the Periodic Table. DNA and RNA are built only by: hydrogen, oxygen, carbon, nitrogen and phosphorus (H, O, C, N and P). The polypeptides of proteins consist only of H, O, C, N and S (sulfur). These atoms have the following characteristics: (1) they are few; (2) four of them are the same in nucleic acids and proteins; (3) the six atoms are not spread over the Periodic Table, but are located only on its right side; (4) they are all non-metals; and (5) they actually occupy a "niche" in the periodic system all of them having a low number of protons in their atom nucleus. These properties indicate that the periodicity occurring at the atomic level has participated in the choice of atoms that directed chromosome organization and function (A. Lima-de-Faria 2001).

This is a key phenomenon that demands further investigation, because during all its future life, of over 2.5 billion years, the chromosome could not depart from the canalization imposed by these six atoms.

Mineral participation in chromosome function
Most researchers, who have investigated the origin of life, agree that minerals have played a role in the emergence of its properties such as replication. Even crystals are endowed with the capacity to assemble atoms into copies of an original atomic configuration. R.M. Hazen (2001) has recently discussed this situation stating that at the origin of the cell "critical transformations might not have been possible without the help of minerals acting as containers, scaffolds, templates, catalysts and reactants".

This problem brings us to the question of mineral participation in chromosome function. The function of many key proteins and other basic cell macromolecules is not derived from their long molecular chains but from single atoms. Examples are: iron (hemoglobin), zinc (zinc fingers), cobalt (vitamin B_{12}), magnesium (chlorophyll) and other atoms which are found in minerals. But critical to our understanding of the chromosome are three features. Firstly, there is no genetic code for iron, magnesium or any other of the atoms mentioned above. These atoms are picked up from the environment. Secondly, these atoms are those that possess the basic property that the macromolecule exhibits such as the iron in hemoglobin and the magnesium in chlorophyll. Thirdly, the atoms had their own properties before the cell and the chromosome had arisen, and the incorporation into the macromolecular edifice only amplified the power of their functions. The iron atom already had its catalytic activity before it became part of the hemoglobin molecule (M. Calvin 1983).

The perpetuation of symmetries and their mechanism of formation

There is also another connection with the mineral world that is equally critical. Most living organisms from plants to invertebrates, to humans, show organ or body symmetry, such as the bilateral symmetry of the human body, the five-fold symmetry of flowers, the radial symmetry of starfishes or the left-handed and right-handed symmetry found in organs of vertebrates including humans. This last type of symmetry already existed in the carbon atom before the genetic code was formed. Even preceding the formation of the carbon atom were the elementary particles, such as the neutrino, which occurred in left-handed and right-handed forms (H.R. Pagels 1982). Hence, symmetries have been perpetuated throughout evolution, starting with the emergence of matter and surviving intact as late as the development of the human species (A. Lima-de-Faria 1997, 1998).

Since the 1940s it has been known that there are genes which are responsible for symmetry transformations in plants, such as the change from bilateral to radial symmetry in snapdragons (H. Stubbe 1966). But the genetic code in itself does not contain this information. The change in symmetry, that the gene conveys, is implemented by the final configuration of the atoms in the molecule that happens to intervene in the development of the flower. If we want to understand the mechanism responsible for the formation of symmetries we need to look for the atomic processes that occur at the level of the final shaping of the organism phenotype.

Hence the information from the gene is valuable but only partial, since to understand the final result, we need to discover what atomic processes decide the establishment of that primeval event that is symmetry.

This was just a glimpse at some of the physical processes that may tell us where the chromosome came from. We may have to wait for a new generation of physicists to elucidate the atomic events occurring within the chromosome and the effect of specific atomic configurations on the phenotype of the cell and of the organism.

THE ATOMIC ORIGIN OF THE CHROMOSOME IS EXPECTED TO CONDITION ITS FUTURE EVOLUTION

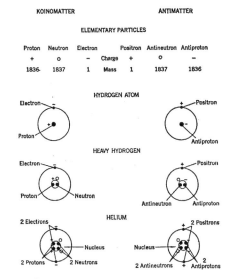

ALFVÉN 1966 -
The particles and antiparticles, known in 1966, which build matter and antimatter.

QUINN and WITHERELL 1998 -
Presently known particles and antiparticles showing their quark content (see page 123).

SANDERSON 1967 - Periodic table of the elements.

FREY-WYSSLING 1945 -
As early as 1945, A. Frey-Wyssling noted that the chemical elements, essential for cell construction and function, occupied a position in the Periodic Table that was non-random. They tend to build a line that goes from Carbon to Argon.

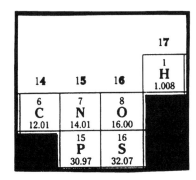

LIMA-DE-FARIA 2001 -
DNA and RNA are built only by: hydrogen, oxygen, carbon, nitrogen, and phosphorus (H, O, C, N and P). The polypeptides of proteins consist only of H, O, C, N and S (sulfur). These atoms are not spread at random over the Periodic Table but are all non-metals and occupy a "niche" in the periodic system.

Where is the chromosome going?

Predictions in science do not represent visions through a crystal ball. Rather, they are the outcome of the assembly of data that have been obtained following a certain coherent scheme.

Prediction is a limited process. One simply uses pieces of information that are already available and combines them in a new way. The view, thus put forward, reflects the evidence and technology available at a given time. Obviously a simple answer is not in sight. But on the basis of the atomic constraints that are becoming evident, there seem to be limitations on the directions that the evolution of the chromosome may take.

The number of human primary genes may be smaller than 1,000
As late as the 1980s it was considered that humans should have at least 200,000 genes. This value was based on the number of proteins estimated to exist in the human body. This figure started to decrease as it was discovered that one DNA sequence could give rise to several proteins and that split genes in humans could consist of as many as 118 exons (A.M. Christiano et al. 1994). Figures of 100,000 and 70,000 started to be mentioned. But it came as a surprise when the sequencing of the human genome led to a gene number of 39,114 (Celera) and 31,780 (public sequence).

The general attitude is that this is far from a definitive value. Actually the number may be much smaller if one considers the primary genes from which the 30,000 may have been derived. The evolution of the sequences of proteins led M.O. Dayhoff et al. (1975) to conclude that all proteins known may belong to fewer than 1,000 distinct superfamilies (groups related by their amino acid sequences). Following this reasoning the number of basic genes in humans could turn out to be of the same magnitude, *i.e.* fewer than 1,000.

The distinction between genetic noise and genetic music
In its permanent molecular activity, and the reshaping of its structure, the chromosome is prone to faulty solutions in its molecular edifice. These take the form of base deletions, base substitutions, accidental rearrangements and other errors that do not agree with its initial construction. These lead to defective mutations of various kinds and may be called genetic noise.

But there is another type of transformations occurring in the chromosome. Thanks to its inherent mechanisms of surveillance, these transformations result in what could be called genetic music. The maintenance of harmony is due to repair mechanisms carried out by specific proteins. The reestablishment of order takes many forms such as: cut-and-patch repair, error-prone repair, mismatch repair, photoreactivating enzyme system, proofreading, recombination repair, SOS response and others (K.-C. Chow 2000, W. Arber 1991 and 2000). Also contributing to the maintenance of the coherence of the chromosome organization are the ordered rearrangements carried out with the help of transposons. These are flanked by inverted repeat sequences and their movement is directed by transposases and resolvases.

As T.A. Brown (1999) has stated "randomness does not apply to all components of the non-coding DNA. In particular, transposable elements and introns have interesting evolutionary histories".

Chromosome length is critical for gene function and location
The preceding question leads us to the problem of chromosome length and to the sensing mechanisms that adjust chromosome behaviour.

Why should the number of nucleotide pairs lying between the centromere and telomere have anything to do with chromosome organization? This question is raised in the light of evidence showing that terminal organizers establish a field of action along the chromosome that determines the location and function of the intervening DNA sequences. A.W. Murray and co-workers (A.W. Murray et al. 1986, A.W. Murray and J.W. Szostak 1987), working with yeast, constructed artificial chromosomes that ranged in length from 42,000 to 150,000 base pairs. The small chromosomes segregated randomly at mitosis but the long ones displayed mitotic stability. From this they concluded that "Increasing the length of artificial chromosomes decreases the rate at which they are lost during mitosis." In the case of these very small chromosomes, it is the telomere-telomere distance that seems to be critical for their function (R.M. Walmsley and T.D. Petes 1985).

Other results disclose that the length of the centromere-telomere region influences the location of DNA sequences. In the very long chromosome arms of the *Liliaceae* and the *Amphibia*, the ribosomal genes tend to show the most irregular patterns of distribution. This is also the case with the very long arms of the chromosomes of maize in which the "knob" sequences diverge from their strict telon location. Moreover, in the smallest arms, which are less than 15 μm (at pachytene), knobs tend to be absent. Thus, there is an optimal centromere-telomere length at which their location is most stable. The same is true of the DNA sequences involved in exchanges in human chromosomes and those occurring in certain Giemsa bands. In the longest arms, they are also found to move away from the telomeres where they have their optimal territory (A. Lima-de-Faria 1980).

The results from yeast and higher organisms point in the direction of a mechanism which would affect the behaviour of DNA sequences located along the centromere-telomere and the telomere-telomere region. The number of nucleotide pairs between these regions would be a critical factor.

Sensing mechanisms

The interactions occurring along the chromosome are expected to be mediated by molecular processes, but the occurrence of sensing mechanisms that would mediate these interactions has been scanty. However, it is now accumulating rapidly. As mentioned above the earliest results were those of F. Ritossa (1973) and of K.D. Tartof (1974). They found that, in *Drosophila*, gene magnification and gene amplification were dependent on a sensing mechanism that was capable of recognizing the number of ribosomal RNA genes present in the chromosome and of introducing the necessary adjustment obliging the gene level to return to the original one.

The protozoan *Trypanosoma* is a parasite which evades the immune system of the host. It can switch on new genes that code for its surface antigens. In this way it rapidly changes the properties of its surface coat. These variable genes are expressed when two events occur: (1) an extra copy of the gene is produced and (2) it is translocated to a telomere region. Only the copy that is located near the telomere produces the messenger RNA of the new glycoprotein. The gene may have different positions along the chromosome but the telomere sequences are necessary for expression (J.E. Donelson and M.J. Turner 1985).

Recently S.K. Evans and V. Lundblad (2000) studying the regulation of telomerase access to the telomere concluded that "DNA-binding proteins are part of a length-sensing mechanism that can discriminate the number of duplex-binding proteins bound to the telomere".

The whole human genome may be packed into a single chromosome

It is true that the chromosome follows its undisturbed path dictated by the laws of physics and chemistry. But we may be able to change it appreciably within limits.

Species of the deer *Muntjak* have 3 and 23 chromosomes (haploid) which result in nearly identical individuals. When we have discovered the rules of chromosome organization, that allow such a transformation, we will be in a position to produce a human being with only 3 (instead of 23) chromosomes which is the present number of the human genome. We may even be able to pack all the human genetic information into a single chromosome, as it happens in the ant *Myrmecia*. In these insects a single chromosome produces an animal nearly indistinguishable from another one with 32 chromosomes (see the detailed description of these chromosome transformations in Part VII). At present the mechanisms, involved in the aggregation as well as the dispersal of chromosome units, escape us completely.

In one hundred years from now our picture of the chromosome may have changed so radically that our present bird's eye view may seem archaic and naive, but that is the expected result of an expanding scientific endeavour.

Bibliography
A list of selected books that have dealt with the chromosome during the period 1870–2001

Detailed references of the works cited in the text may be found in these publications. The list does not pretend to be exhaustive but consists mainly of works pertinent to the field of chromosome research.

1870–1899

1870 Waldeyer, W. Eirstock und Ei. Leipzig.
1880 Hanstein, J. Das Protoplasma als Träger der pflänzlichen und thierischen Lebensverrichtungen. Heidelberg.
1882 Flemming, W. Zellsubstanz, Kern und Zelltheilung. Leipzig.
1883 Roux, W. Die Bedeutung der Kerntheilungsfiguren. Engelmann, Leipzig.
1885 Weismann, A. Die Continuität des Keimplasmas als Grundlage einer Theorie der Vererbung. Jena.
1886 Berthold, G. Studien über Protoplasma-mechanik. Leipzig.
1889 De Vries, H. Intracellular Pangenesis. Jena.
1892 Hertwig, O. Ältere und neuere Entwicklungstheorieen. Berlin.
1892 Wiesner, J. Die Elementarstructur und das Wachstum der lebenden Substanz. Wien.
1892 Bütschli, O. Untersuchungen über mikroscopische Schäume und das Protoplasma. Leipzig. Eng. Trans., E.A. Minchin, London.
1893 Weismann, A. The Germ-plasm. Eng. Tr. by W.N. Parker and H. Ronnfeldt, New York.
1894 Driesch, H. Analytische Theorie der organischen Entwicklung. Leipzig.
1899 Correns, C. Untersuchungen über die Vermehrung der Laubmoose durch Brutorgane und Stecklinge. Jena.

1900–1909

1903 Strasburger, E. Text-book of Botany (Strasburger, Noll, Schenk, and Schimper). 2nd Eng. Edition from 5th German Edition.
1904 Boveri, Th. Ergebnisse über die Konstitution der chromatischen Substanz des Zellkerns. Jena.
1904 Gurwitsch, A. Morphologie und Biologie der Zelle. Jena.
1905 Boveri, Th. Zellenstudien V. Ueber die Abhängigkeit der Kerngrösse und Zellenzahl der Ausgangzellen. Jena.
1906 Hertwig, R. Eireife und Befruchtung. In O. Hertwig Handbuch d. Vergl. und Experim. Entwickelungslehre I. Fischer, Jena.
1907 Morgan, T.H. Experimental Zoölogy. Macmillan, New York.
1908 Brachet, A. L'hérédité dans l'œuf. Revue des Idées.
1909 Bateson, W. Mendel's principles of heredity. University Press, Cambridge.

1910–1919

1912 Meyer, A. Die Zelle der Bakterien. Jena.
1912 Kossel, A. The Proteins. Herter Lecture. Johns Hopkins Hospital Bulletin, Baltimore.
1913 Weismann, A. Vorträge über Deszendenz Theorie. 3rd Auflage. Fischer, Jena.
1913 Morgan, T.H. Heredity and Sex. Columbia University Press, N.Y.
1913 Warburg, O.H. Ueber die Wirksamkeit der Strukturen auf die chemischen Vorgänge in Zellen. Jena.
1913 Molisch, H. Mikrochemie der Pflanze. Jena.
1913 Correns, C. and Goldschmidt, R. Die Vererbung und Bestimmung des Geschlechtes. Berlin.
1913 Goldschmidt, R. Einführung in die Vererbungswissenschaft. 2 Aufl. Leipzig.
1913 Loeb, J. Artificial Parthenogenesis and Fertilization. University Chicago Press.
1913 Caullery, M. Les problèmes de la sexualité. Flammarion, Paris.
1914 Baur, E. Einführung in die experimentelle Vererbungslehre. Borntraeger, Berlin.
1914 Coulter, J.M. Evolution of Sex in Plants. Chicago.
1914 Boveri, Th. Zur Frage der Entstehung maligner Tumoren. Fischer, Jena.
1914 Doncaster, L. The Determination of Sex. Cambridge University Press, Cambridge.
1915 Buchner, P. Praktikum der Zellenlehre. Bornträger, Berlin.
1915 Child, C.M. Individuality in Organisms. Chicago.
1915 Gates, R.R. The Mutation Factor in Evolution, with Particular Reference to Œnothera. London.
1916 Loeb, J. The Organism as a Whole. New York.
1917 Morgan, T.H. The Theory of the Gene. A.N., LI.
1917 Thompson, D'Arcy W. On Growth and Form. Cambridge.
1917 Hartmann, M. and Schillings, C. Die Pathogenen Protozoa. Springer, Leipzig.
1919 East, E.M. and Jones, D.F. Inbreeding and outbreeding. J.B. Lippincott, Philadelphia.
1919 Lillie, F.R. Problems of Fertilization. University Chicago Press.
1919 Morgan, T.H. The Physical Basis of Heredity. Philadelphia.

1920–1929

1920 Doncaster, L. An Introduction to the Study of Cytology. Cambridge University Press.

| 1920 | Agar, W.E. Cytology with special Reference to the Metazoan Nucleus. Macmillan and Co., London.
| 1920 | Winkler, H. Verbreitung und Ursache der Parthenogenesis im Pflanzen- und Tierreiche. Jena.
| 1920 | Goldschmidt, R. Mechanismus und Physiologie der Geschlechtsbestimmung. Bornträger, Berlin.
| 1921 | Sharp, L.W. An Introduction to Cytology. McGraw-Hill, New York.
| 1922 | Hatschek, E. An Introduction to the Physics and Chemistry of Colloids. 4th Ed. Philadelphia.
| 1923 | Bayliss, W.M. The Colloidal State. London.
| 1924 | McClung, C.E. The Chromosome Theory of Heredity. In General Cytology, Chicago.
| 1925 | Wilson, E.B. The Cell in Development and Heredity. Third edn. The Macmillan Company.
| 1927 | Crew, F.A.E. The genetics of sexuality in animals. Cambridge University Press.
| 1927 | Babcock, E.B. and Clausen, R.E. Genetics in relation to agriculture. McGraw-Hill, New York.
| 1928 | Schrader, F. The sex chromosomes. Borntraeger, Berlin.
| 1929 | Roberts, H.F. Plant hybridization before Mendel. University Press, Princeton.
| 1929 | Renner, O. Artbastarde bei Pflanzen. Borntraeger, Berlin.

1930–1939

| 1930 | Belling, J. The Use of the Microscope. McGraw-Hill, New York.
| 1930 | Fisher, R.A. The genetical theory of natural selection. Clarendon Press, Oxford.
| 1932 | Darlington, C.D. Recent Advances in Cytology. J. & A. Churchill, London.
| 1932 | Haldane, J.B.S. The causes of evolution. Harper and Brothers, New York.
| 1933 | Matsuura, H. A bibliographical monograph on plant genetics. Hokkaido University, Sapporo.
| 1933 | Stern, C. Faktorenkoppelung und Faktoren-austausch. Handbuch Vererbungswiss. I, H.
| 1934 | Sharp, L.W. Introduction to cytology. McGraw-Hill, New York.
| 1934 | Morgan, T.H. Embryology and genetics. Columbia University Press, New York.
| 1935 | Emerson, R.A., Beadle, G.W. and Fraser, A.C. A summary of linkage studies in maize. Memoir 180, Cornell University Agric. Exper. Sta.
| 1935 | Koltzoff, N.K. Physiologie du Dévelopement et Génétique. Hermann, Paris.
| 1936 | Hertwig, P. Artbastarde bei Tieren. Handbuch Vererbungswiss., 2, B.
| 1936 | Duggar, B.M. et al. Biological effects of radiation. 2 vols., McGraw-Hill Co., New York.
| 1937 | Dobzhansky, T. Genetics and the origin of species. Columbia University Press, New York.
| 1937 | Cook, R. A chronology of genetics. U.S. Dept. Agric. Yearbook.
| 1937 | Newman, H.H., Freeman, F.N. and Holzinger, K.J. Twins: A study of heredity and environment. University Chicago Press, Chicago.
| 1937 | Timoféeff-Ressovsky, N.W. Experimentelle Mutationsforschung in der Vererbungslehre. T. Steinkopf, Dresden.
| 1938 | Goldschmidt, R. Physiological genetics. McGraw-Hill Co., New York.
| 1938 | Mather, K. The measurement of linkage in heredity. Methuen and Co., London.
| 1938 | Stubbe, H. Genmutation. Handbuch Vererbungswiss. 2, F.
| 1939 | Sturtevant, A.H. and Beadle, G.W. An Introduction to genetics. Saunders, Philadelphia.
| 1939 | Sinnott, E.W. and Dunn, L.C. Principles of Genetics. Third Edition. McGraw-Hill, New York.

1940–1949

| 1940 | Clausen, J., Keck, D.D. and Hiesey, W.M. Experimental studies on the nature of species I. The effect of varied environments on western North-American plants. Carnegie Instit. Wash. Publ. No 520.
| 1941 | Boyd, E. Outline of physical growth and development. Burgess Pub. Co.
| 1941 | Guilliermond, A. The cytoplasm of the plant cell. Chronica Botanica, Waltham.
| 1942 | Darlington, C.D. and La Cour, L. The handling of chromosomes. G. Allen, London.
| 1942 | Guyénot, E. L'Hérédité. G. Doin, Paris.
| 1943 | Sharp, L.W. Fundamentals of Cytology. McGraw-Hill, New York.
| 1944 | Schrader, F. Mitosis. Columbia University Press, New York.
| 1944 | Schrödinger, E. What is Life? Cambridge University Press.
| 1945 | White, M.J.D. Animal Cytology and Evolution. Cambridge University Press.
| 1945 | Darlington, C.D. and Janaki-Ammal, E.K. Chromosome atlas of cultivated plants. G. Allen, London.
| 1946 | Fischer, A. Biology of tissue cells. University Press, Cambridge.
| 1946 | Darlington, C.D. The Evolution of Genetic Systems. Cambridge University Press.
| 1947 | Eames, A.J. and MacDaniels, L.H. An introduction to plant anatomy. McGraw, N.Y.
| 1947 | Babcock, E.B. The Genus Crepis, I and II, University Calif. Publ. Bot. 21, 22.
| 1948 | Frey-Wyssling, A. Submicroscopic morphology of protoplasm and its derivatives. Elsevier, N.Y.
| 1949 | Darlington, C.D. and Mather, K. The elements of genetics. Allen and Unwin, London.

1950–1959

| 1950 | Caspersson, T.O. Cell growth and cell function. Chapman and Hall, London.
| 1950 | Waddington, C.H. An Introduction to Modern Genetics. Allen & Unwin, London.
| 1950 | Davidson, J.N. The Biochemistry of the Nucleic Acids. Methuen, London.
| 1950 | White, M.J.D. The Chromosomes. Methuen, London.
| 1950 | Stebbins, G.L. Variation and evolution in plants. Columbia University Press, N.Y.
| 1951 | Clausen, J. Stages in the evolution of plant species. Cornell University Press, New York.

1951 Tischler, G. Allgemeine Pflanzenkaryologie. 2. Hälfte: Kernteilung und Kernverschmelzung. Handbuch der Pflanzenanatomie. Borntraeger.
1951 Dobzhansky, T. Genetics and the origin of species. Columbia University Press, N.Y.
1951 Catcheside, D.G. The Genetics of Micro-Organisms. Pitman, London.
1951 Makino, S. An atlas of the chromosome numbers in animals. Iowa State College Press, Ames.
1952 Patterson, J.T. and Stone, W.W. Evolution in the genus Drosophila. Macmillan, New York.
1952 Hughes, A. The Mitotic Cycle. Butterworths, London.
1952 Manton, I. Problems of cytology and evolution in the Pteridophyta. Cambridge University Press.
1953 Schrader, F. Mitosis: the movement of chromosomes in cell division. Columbia University Press, N.Y.
1953 Ephrussi, B. Nucleo-Cytoplasmic Relations in Micro-Organisms. Clarendon Press, Oxford.
1953 Danielli, J.F. Cytochemistry. A Critical Approach. Wiley, New York.
1954 White, M.J.D. Animal cytology and evolution. Cambridge University Press.
1954 Haldane, J.B.S. The biochemistry of genetics. G. Allen, London.
1955 Goldschmidt, R.B. Theoretical genetics. University of California Press, Berkeley.
1955 Bisset, K.A. The cytology and life-history of bacteria. 2nd ed. Williams & Wilkins, Baltimore.
1955 Darlington, C.D. and Wylie, A.P. Chromosome atlas of flowering plants. Allen and Unwin, London.
1955 Wagner, R.P. and Mitchell, H.K. Genetics and metabolism. Chapman and Hall, London.
1956 Waddington, C.H. Principles of embryology. Allen and Unwin, London.
1956 Darlington, C.D. Chromosome Botany. Allen and Unwin, London.
1957 Beatty, R.A. Parthenogenesis and polyploidy in mammalian development. Cambridge University Press.
1957 Waddington, C.H. The strategy of the genes. Allen and Unwin, London.
1957 Brachet, J. Biochemical Cytology. Academic Press, New York.
1958 Engström, A. and Finean, J.B. Biological Ultrastructure. Academic Press, New York.
1959 Pontecorvo, G. Trends in genetic analysis. Oxford University Press.
1959 Anfinsen, C.B. The molecular basis of evolution. Wiley, New York.
1959 Brachet, J. and Mirsky, A.E. (eds.) The cell, vols. 1–6, 1959–1964. Academic Press, London.

1960–1969

1960 Swanson, C.P. Cytology and Cytogenetics. Macmillan, London.
1961 Jacob, F. and Wollman, E.L. Sexuality and the Genetics of Bacteria. Academic Press, New York.
1961 Oparin, A.I. Life. Its Nature, Origin and Development. Oliver and Boyd, Edinburgh.
1962 Fraenkel-Conrat, H. Design and Function at the Threshold of Life: The Viruses. Academic Press, N.Y.
1962 Asimov, I. The Genetic Code. Signet Science Library, New York.
1962 Perutz, M.F. Proteins and Nucleic Acids. Structure and Function. Elsevier, Amsterdam.
1963 Lewis, K.R. and John, B. Chromosome Marker. J. & A. Churchill, London.
1963 Ingram, V.M. The Hemoglobins in Genetics and Evolution. Columbia University Press, New York.
1963 Wolff, S. (Edit.) Radiation-Induced Chromosome Aberrations. Columbia University Press, New York.
1964 Hayes, W. The Genetics of Bacteria and their Viruses. Blackwell, Oxford.
1964 Locke, M. (Edit.) The Role of Chromosomes in Development. Academic Press, New York.
1964 Bonner, J. and Ts'o, P. (Edit.) The Nucleohistones. Holden-Day, San Francisco.
1965 John, B. and Lewis, K.R. The Meiotic System. Protoplasmatologia. Band VI. Springer-Verlag, Wien, New York.
1965 Sharma, A.K. and Sharma, A. Chromosome Techniques. Butterworths, London.
1965 Bonner, J. The Molecular Biology of Development. Clarendon Press, Oxford.
1965 Watson, J.D. Molecular Biology of the Gene. W.A. Benjamin, New York.
1965 Sturtevant, A.H. A History of Genetics. Harper & Row, New York.
1965 Fox, S.W. (Edit.) The Origins of Prebiological Systems. Academic Press, New York.
1966 Carlson, E.A. The Gene: A Critical History. Saunders, Philadelphia.
1966 Ingram, V.M. The Biosynthesis of Macromolecules. Benjamin, New York.
1967 Ohno, S. Sex chromosomes and Sex-Linked Genes. Springer-Verlag, Berlin, Heidelberg.
1968 Rieger, R., Michaelis, A. and Green, M.M. A Glossary of Genetics and Cytogenetics. Third Edition. Springer-Verlag, Berlin.
1968 Hayes, W. The Genetics of Bacteria and their Viruses. Blackwell, Oxford.
1969 Lima-de-Faria, A. (Edit.) Handbook of Molecular Cytology. North Holland, Amsterdam.
1969 Ycas, M. The Biological Code. North-Holland, Amsterdam.
1969 Benirschke, K. Comparative Mammalian Cytogenetics. Springer-Verlag, Berlin.

1970–1979

1970 Ohno, S. Evolution by Gene Duplication. Springer-Verlag, Berlin.
1970 Busch, H. and Smetana, K. The Nucleolus. Academic Press, New York.
1970 DuPraw, E.J. DNA and Chromosomes. Holt, Rinehart and Winston.
1970 Beckwith, J.R. and Zipser, D. (Edit.) The Lactose Operon. Cold Spring Harbor Lab., N.Y.
1971 Stebbins, G.L. Chromosomal Evolution in Higher Plants. Addison-Wesley, London.
1971 Stent, G.S. Molecular Genetics. Freeman, San Francisco.

1971 Hamerton, J.L. Human Cytogenetics I and II. Academic Press, New York.
1971 Hershey, A.D. (Edit.) The Bacteriophage Lambda. Cold Spring Harbor Lab., N.Y.
1971 Rutten, M.G. The Origin of Life. Elsevier, Amsterdam.
1972 Sirlin, J.L. Biology of RNA. Academic Press, New York.
1972 Ursprung, H. (Edit.) Nucleic Acid Hybridization in the Study of Cell Differentiation. Springer-Verlag, Berlin.
1973 Chiarelli, A.B. and Capanna, E. Cytotaxonomy and Vertebrate Evolution. Academic Press, London.
1973 Stern, C. Principles of Human Genetics. W.H. Freeman, San Francisco.
1973 Whitehouse, H.L.K. Towards an Understanding of the Mechanism of Heredity. Edward Arnold, London.
1973 Orgel, L.E. The Origins of Life. Chapman and Hall, London.
1974 Gurdon, J.B. The Control of Gene Expression in Animal Development. Clarendon Press, Oxford.
1974 Lewin, B. Gene Expression. Vol. 1 (Bacterial Genomes). John Wiley, London.
1974 Lewin, B. Gene Expression, Vol. 2. Eukaryotic Chromosomes. John Wiley, London.
1974 Busch, H. (Edit.) The Cell Nucleus. Academic Press, New York. Volumes 1 to 7, 1974 to 1979.
1974 German, J. Chromosomes and Cancer. John Wiley, New York.
1975 Harris, H. The Principles of Human Biochemical Genetics. North-Holland, Amsterdam.
1975 Makino, S. Human Chromosomes. Igaku Shoin Ltd., Tokyo.
1976 Nagl, W. Zellkern und Zellzyklen. Eugen Ulmer, Stuttgart.
1976 Ringertz, N.R. and Savage, R.E. Cell Hybrids. Academic Press, New York.
1976 Bodmer, W.F. and Cavalli-Sforza, L.L. Genetics, Evolution, and Man. W.H. Freeman, San Francisco.
1976 Davidson, E.H. Gene Activity in Early Development. Academic Press, New York.
1976 Adams, R.L.P., Burdon, R.H., Campbell, A.M. and Smellie, R.M.S. Davidson's The Biochemistry of the Nucleic Acids. Chapman and Hall, London.
1977 Lewin, B. Gene Expression. Vol. 3 (Plasmids & Phages). John Wiley, New York.
1977 Kihlman, B.A. Caffeine and Chromosomes. Elsevier, Amsterdam.
1977 D'Amato, F. Nuclear Cytology in Relation to Development. Cambridge University Press, London.
1978 Denhardt, D.T., Dressler, D. and Ray, D.S. (Edit.) The Single-Stranded DNA Phages. Cold Spring Harbor Lab. N.Y.
1978 Bostock, C.J. and Sumner, A.T. The Eukaryotic Chromosome. North-Holland, Amsterdam.
1978 Nagl, W. Endopolyploidy and Polyteny in Differentiation and Evolution. North-Holland, Amsterdam.
1978 White, M.J.D. Modes of Speciation. Freeman, San Francisco.
1978 Miller, J.H. and Reznikoff, W.S. (Edit.) The Operon. Cold Spring Harbor Lab., N.Y.
1979 Dyer, A.F. Investigating Chromosomes. Edward Arnold, London.
1979 Hsu, T.C. Human and Mammalian Cytogenetics. An Historical Perspective. Springer-Verlag, New York.
1979 Seuánez, H.N. The Phylogeny of Human Chromosomes. Springer-Verlag, Berlin.
1979 Broda, P. Plasmids. W.H. Freeman and Company, San Francisco.
1979 Ohno, S. Major Sex-Determining Genes. Springer-Verlag, Berlin.

1980–1989

1980 Kornberg, A. DNA Replication. W.H. Freeman and Company, San Francisco.
1980 Schulz-Schaeffer, J. Cytogenetics. Plants – Animals – Humans. Springer-Verlag, New York.
1980 Therman, E. Human Chromosomes. Structure, Behavior, Effects. Springer-Verlag, New York.
1980 Sandberg, A.A. The Chromosomes in Human Cancer and Leukemia. Elsevier, New York.
1980 Ferguson, A. Biochemical Systematics and Evolution. Blackie, Glasgow.
1980 Old, R.W. and Primrose, S.B. Principles of Gene Manipulation. Blackwell, Oxford.
1981 Bradbury, E.M., MacLean, D.N. and Matthews, H. DNA, Chromatin and Chromosomes. Blackwell, Oxford.
1981 Swanson, C.P., Merz, T. and Young, W.J. Cytogenetics. The Chromosome in Division, Inheritance, and Evolution. Prentice-Hall, Englewood Cliffs, N.J. Second Edn.
1981 Margulis, L. Symbiosis in Cell Evolution. Freeman, San Francisco.
1982 Tamarin, R. and Leavitt, R.W. Principles of Genetics. Wm. C. Brown Publishers.
1983 Alberts, B., Bray, D., Lewis, J., Raff, M., Roberts, K. and Watson, J.D. Molecular Biology of the Cell. Garland Publishing, New York.
1983 Lima-de-Faria, A. Molecular Evolution and Organization of the Chromosome. Elsevier, Amsterdam.
1983 Mitelman, F. Catalogue of Chromosome Aberrations in Cancer. Karger.
1983 Lewin, B. Genes. John Wiley, New York.
1984 Connor, M. and Ferguson-Smith, M. Essential Medical Genetics. Blackwell Science.
1984 Glover, D.M. Gene Cloning. Chapman and Hall, London.
1984 de Duve, C. A Guided Tour of the Living Cell. Volume One. Scientific American Library, New York.
1984 de Duve, C. A Guided Tour of the Living Cell. Volume Two. Scientific American Library, New York.
1984 Taylor, J.H. DNA Methylation and Cellular Differentiation. Springer-Verlag, New York.
1985 De Pomerai, D. From Gene to Animal. An Introduction to the Molecular Biology of Animal Development. Cambridge University Press, London.
1985 Friedberg, E.C. DNA Repair. W.H. Freeman, New York.

1985 Snyder, L.A., Freifelder, D. and Hartl, D.L. General Genetics. Jones and Bartlett, Boston.
1986 Lodish, H., Baltimore, D., Berk, A., Zipursky, S.L., Matsudaira, P. and Darnell, J. Molecular Cell Biology. W.H. Freeman, New York.
1986 Pritchard, D.J. Foundations of Developmental Genetics. Taylor & Francis, London.
1986 Spirin, A.S. Ribosome Structure and Protein Biosynthesis. Benjamin/Cummings, Reading, Massachusetts.
1987 Heim, S. and Mitelman, F. Cancer Cytogenetics. Alan R. Liss., New York.
1987 Watson, J.D., Hopkins, N.H., Roberts, J.W., Steitz, J.A. and Weiner, A.M. Molecular Biology of the Gene. Benjamin/Cummings, Menlo Park, California.
1988 Fristrom, J.W. and Clegg, M.T. Principles of Genetics. W.H. Freeman, New York.
1988 Lima-de-Faria, A. Evolution without Selection. Form and Function by Autoevolution. Elsevier, Amsterdam.
1989 Pugsley, A.P. Protein Targeting. Academic Press, San Diego.

1990–2001

1990 Gelehrter, T.D. and Collins, F.S. Genetics. Williams & Wilkins, Baltimore.
1991 Singer, M. and Berg, P. Genes & Genomes. Blackwell, Oxford.
1991 Branden, C. and Tooze, J. Introduction to Protein Structure. Garland Publishing, New York.
1992 Lawrence, P.A. The Making of a Fly. The Genetics of Animal Design. Blackwell, Oxford.
1993 Griffiths, A.J.F., Miller, J.H., Suzuki, D.T., Lewontin, R.C. and Gelbart, W.M. An Introduction to Genetic Analysis. W.H. Freeman, New York.
1993 Murray, A. and Hunt, T. The Cell Cycle: An Introduction. W.H. Freeman, New York.
1993 Varmus, H. and Weinberg, R.A. Genes and the Biology of Cancer. Scientific American Library, New York.
1994 Lewin, B. Genes V. Oxford University Press.
1995 Lima-de-Faria, A. Biological Periodicity. Its Molecular Mechanism and Evolutionary Implications. JAI Press Inc., Greenwich, Connecticut.
1995 Kipling, D. The Telomere. Oxford University Press.
1995 Weaver, R.F. and Hedrick, P.W. Basic Genetics. Wm. C. Brown Publishers, Dubuque, Iowa.
1995 Jorde, L.B., Carey, J.C. and White, R.L. Medical Genetics. Mosby, St. Louis.
1996 Raff, R.A. The Shape of Life. Genes, Development, and the Evolution of Animal Form. University of Chicago Press, Chicago.
1996 Gall, J.G. A Pictorial History – Views of the Cell. The American Society for Cell Biology.
1997 King, R.C. and Stansfield, W.D. A Dictionary of Genetics. Oxford University Press.
1997 Choo, K.H.A. The Centromere. Oxford University Press.
1998 McKusick, V.A. Mendelian Inheritance in Man. A Catalog of Human Genes and Genetic Disorders. Twelfth Edition. Volume 1. John Hopkins University Press, Baltimore.
1999 Brown, T.A. Genomes. Bios Scientific Publishers, Oxford.
2000 Lewin, B. Genes VII. Oxford University Press.
2000 Gilbert, S.F. Developmental Biology. Sixth Edition. Sinauer Associates, Sunderland, Massachusetts.
2000 Klug, W.S. and Cummings, M.R. Concepts of Genetics. Sixth Edition. Prentice Hall, Upper Saddle River, New Jersey.
2000 Keller, E.F. The Century of the Gene. Harvard University Press, Cambridge, Massachusetts.
2001 Morange, M. The Misunderstood Gene. Harvard University Press, Cambridge, Massachusetts.

References
Cited works between 1990 and 2001

Akey, C.W. and Radermacher, M. 1993. Architecture of the Xenopus nuclear pore complex revealed by three-dimensional cryo-electron microscopy. Journal Cell Biology 122: 1–19.

Alberts, B. et al. 1994. Molecular Biology of the Cell. Garland Publishing, Inc., New York.

Allshire, R. and Bickmore, W. 2000. Pausing for thought on the boundaries of imprinting. Cell 102: 705–708.

Anderson, W.F. 1990. Whither goest thou, gene therapy? Hum Gene Ther 1(3): 227–228.

Anderson, W.F. et al. 1990. The ADA human gene therapy clinical protocol. Hum Gene Ther 1(3): 331–362.

Arber, W. 1991. Elements in microbial evolution. J Mol Evol 33: 4–12.

Arber, W. 1997. The influence of genetic and environmental factors on biological evolution. In: Pontifical Academy of Sciences, Plenary Session on 'The origin and early evolution of life' (Part 1). Commentarii 4, Vatican City: 81–100.

Arber, W. 1999. Involvement of gene products in bacterial evolution. Ann. New York Acad Sci 851: 36–44.

Arber, W. 2000. Genetic variation: molecular mechanisms and impact on microbial evolution. FEMS Microbiol Rev 24: 1–7.

Arms, K. and Camp, P.S. 1995. Biology. Saunders College Publishing and Harcourt Brace College Publishers, New York.

Baker, S.J. et al. 1990. Suppression of human colorectal-carcinoma cell-growth by wild-type-P53. Science 249: 912–915.

Bartel, D.P. and Unrau, P.J. 1999. Constructing an RNA world. TIBS 24(12): M9–12.

Bell, A.C. et al. 2001. Insulators and boundaries: versatile regulatory elements in the eukaryotic genome. Science 291: 447–497.

Bertolotto, C.E.V. et al. 2001. Banding patterns, multiple sex chromosome system and localization of telomeric $(TTAGGG)_n$ sequences by FISH on two species of Polychrus (Squamata, Polychrotidae). Caryologia 54(3): 217–226.

Bhattacharyya, M.K. et al. 1990. The wrinkled-seed character of pea described by Mendel is caused by a transposon-like insertion in a gene encoding starch-branching enzyme. Cell 60(1): 115–122.

Blackburn, E.H. 2000. The end of the (DNA) line. Nature Structural Biology 7(10): 847–849.

Blum, B. et al. 1990. A model for RNA editing in kinetoplastid mitochondria: "guide" RNA molecules transcribed from maxicircle DNA provide the edited information. Cell 60(2): 189–198.

Bollag, R.J. et al. 1994. An ancient family of embryonically expressed mouse genes sharing a conserved protein motif with the T-locus. Nature Genetics 7(3): 383–389.

Borie, N. et al. 2000. Developmental expression of the 412 retrotransposon in natural populations of *D. melanogaster* and *D. simulans*. Genet Res, Camb 76: 217–226.

Bork, P. and Copley, R. 2001. Filling in the gaps. Nature 409: 818–820.

Bradley, R.D. and Wichman, H.A. 1994. Rapidly evolving repetitive DNAs in a conservative genome: a test of factors that affect chromosomal evolution. Chromosome Research 2: 354–360.

Brown, T.A. 1999. Genomes. Bios Scientific Publishers.

de Bruin, D. et al. 2000. Telomere folding is required for the stable maintenance of telomere position effects in yeast. Molecular and Cellular Biology 20(21): 7991–8000.

Brusca, R.C. and Brusca, G.J. 1990. Invertebrates. Sinauer Associates, Inc. Publishers Sunderland, Massachusetts.

Bult, C.J. et al. 1996. Complete genome sequence of the methanogenic archaeon, Methanococcus jannaschii. Science 273: 1058–1073.

Caprara, M.G. and Nilsen, T.W. 2000. RNA: Versatility in form and function. Nature Structural Biology 7(10): 831–833.

Cavalier-Smith, T. 1991. The evolution of prokaryotic and eukaryotic cells. In: Fundamentals of Medical Cell Biology, Volume 1. Evolutionary Biology. JAI Press Inc.: 217–272.

Cedergren, R. and Miramontes, P. 1996. The puzzling origin of the genetic code. Trends Biochem Sci 21: 199–200.

Chen, C.W. 1996. Complications and implications of linear bacterial chromosomes. TIG 12(5): 192–196.

Cheng, E.H.-Y.A. et al. 2001. BCL-2, BCL-XL sequester BH3 domain-only molecules preventing BAX- and BAK-mediated mitochondrial apoptosis. Molecular Cell 8: 705–711.

Chikashige, Y. et al. 1994. Telomere-led premeiotic chromosome movement in fission yeast. Science 264: 270–273.

Choo, K.H.A. 1997. The Centromere. Oxford University Press, Oxford.

Chow, K.-C. 2000. Hsp70 (DnaK) – an evolution facilitator? TIG 16(11): 484–485.

Chowdhary, B.P. et al. 1998. Emerging patterns of comparative genome organization in some mammalian species as revealed by Zoo-FISH. Genome Research 8: 577–589.

Christiano, A.M. et al. 1994. Structural Organization of the human type VII collagen gene (COL7A1), composed of more exons than any previously characterized gene. Genomics 21: 169–179.

Culbertson, M.R. 1999. RNA surveillance. Unforeseen consequences for gene expression, inherited genetic disorders and cancer. TIG 15: 74–75.

Doxsey, S.J. et al. 1994. Pericentrin, a highly conserved centrosome protein involved in microtubule organization. Cell 76: 639–650.

Drawid, A. et al. 2000. Genome-wide analysis relating expression level with protein subcellular localization. TIG 16(10): 426–428.

Dubnau, J. and Struhl, G. 1996. RNA recognition and translational regulation by a homeodomain protein. Nature 379: 694–699.

Dujon, B. 1996 The yeast genome project: what did we learn?. TIG 12(7): 263–270.

Dujon, B. et al. 1994. Complete DNA-sequence of yeast chromosome-XI. Nature 369: 371–378.

Dunham, I. 2000. Genomics – the new rock and roll? TIG 16(10): 456–460.

Dunham, I. et al. 1999. The DNA sequence of human chromosome 22. Nature 402: 489–495.

Evans, S.K. and Lundblad, V. 2000. Positive and negative regulation of telomerase access to the telomere. Journal of Cell Science 113: 3357–3364.

Fleischmann, R.D. et al. 1995. Whole-genome random sequencing and assembly of Haemophilus-influenzae RD. Science 269: 496–512.

Fothergill-Gilmore, L.A. 1991. The evolution of RNA and protein as catalysts. In: Fundamentals of Medical Cell Biology, Volume 1. Evolutionary Biology. JAI Press Inc.: 163–188.

Fraser, C.M. et al. 1995. The minimal gene complement of mycoplasma-genitalium. Science 270: 397–403.

Frönicke, L. et al. 1997. Segmental homology among cattle (*Bos taurus*), Indian muntjac (*Muntiacus muntjak vaginalis*), and Chinese muntjac (*M. reevesi*) karyotypes. Cytogenet. Cell Genet. 77: 223–227.

Frönicke, L. and Scherthan, H. 1997. Zoo-fluorescence in situ hybridization analysis of human and Indian muntjac karyotypes (*Muntiacus muntjak vaginalis*) reveals satellite DNA clusters at the margins of conserved syntenic segments. Chromosome Research 5: 254–261.

Gineitis, A.A. et al. 2000. Human sperm telomere-binding complex involves histone H2B and secures telomere membrane attachment. The Journal of Cell Biology 151(7): 1591–1597.

Goldman, R.D. 1995. Photograph included in Fig. 23.58 of 1995 H. Lodish et al. Molecular Cell Biology. Scientific American Books, W.H. Freeman and Company, New York.

Goldstein, J.L. 2001. Laskers for 2001: Knockout mice and test-tube babies. Nature Medicine 7(10): 1079–1080.

Griffiths, A.J.F. et al. 1993. An Introduction to Genetic Analysis. W.H. Freeman and Company, New York.

Gurney, M.E. et al. 1994. Motor-neuron degeneration in mice that express a human Cu, Zn superoxide-dismutase mutation. Science 264: 1772–1775.

Haber, J.E. 1999. DNA recombination: the replication connection. TIBS 24 – July 1999: 271–275.

Halder, G. et al. 1995. Induction of ectopic eyes by targeted expression of the eyeless gene in *Drosophila*. Science 267: 1788–1792.

Harrington, J. J. et al. 1997. Formation of *de novo* centromeres and construction of first-generation human artificial microchromosomes. Nature Genetics (15 April): 345–354.

Hazen, R.M. 2001. Life's rocky start. Scientific American April 2001: 77–85.

Hendrix, R.W. 1999. Evolution: The long evolutionary reach of viruses. Current Biology 9: R914–R917.

Henikoff, S. 1992. Position effect and related phenomena. Curr Opin Genet Dev 2: 907–912.

Horai, S. et al. 1995. Recent African origin of modern humans revealed by complete sequences of hominoid mitochondrial DNAs. P.N.A.S. 92(2): 532–536.

Hudson, T.J. et al. 2001. A radiation hybrid map of mouse genes. Nature Genetics 29: 201–205.

Hughes, J. et al. 1995. The polycystic kidney-disease-1 (PKD1) gene encodes a novel protein with multiple cell recognition domains. Nature Genetics 10(2): 151–160.

Ijdo, J.W. et al. 1991. Origin of human chromosome-2 – An ancestral telomere telomere fusion. P.N.A.S. 88: 9051–9055.

Imai, H.T. et al. 1990. Notes on the remarkable karyology of the primitive ant *Nothomyrmecia macrops*, and of the related genus Myrmecia (Hymenoptera: Formicidae). Psyche 97: 133–140.

International Human Genome Sequencing Consortium. 2001. Initial sequencing and analysis of the human genome. Nature 409: 860–921.

Jones, P. 2001. Smart proteins. New Scientist 17 March 2001: 1–3.

Kandasamy, M.K. and Meagher, R.B. 1999. Actin-organelle interaction: Association with chloroplast in Arabidopsis leaf mesophyll cells. Cell Motility and the Cytoskeleton 44: 110–118.

Kerrebrock, A.W. et al. 1995. MEI-S332, a *Drosophila* protein required for sister-chromatid cohesion, can localize to meiotic centromere regions. Cell 83(2): 247–256.

Kim, N.W. et al. 1994. Specific association of human telomerase activity with immortal cells and cancer. Science 266: 2011–2015.

Kipling, D. 1995. The Telomere. Oxford University Press, Oxford.

Knight, R.D. et al. 1999. Selection, history and chemistry: the three faces of the genetic code. TIBS 24: 241–247.

Knighton, D.R. et al. 1991. Crystal-structure of the catalytic subunit of cyclic adenosine-monophosphate dependent protein-kinase. Science 253: 407–414.

Lawrence, P.A. 1992. The Making of a Fly. The Genetics of Animal Design. Blackwell Scientific Publications, Oxford.

Lazarides, E. 1994. Photograph included in Fig. 2.12 of B. Lewin Genes V. Oxford University Press, Oxford 1994.

Lee, C. et al. 1993. Interstitial localization of telomeric DNA sequences in the Indian muntjac chromosomes: further evidence for tandem chromosome fusions in the karyotypic evolution of the Asian muntjacs. Cytogenet. Cell Genet. 63: 156–159.

Lewin, B. 1994. Genes V. Oxford University Press, Oxford.

Li, W.-H. and Graur, D. 1991. Fundamentals of Molecular Evolution. Sinauer Associates, Inc., Sunderland, Massachusetts.

Liljas, A. and Laurberg, M. 2000. A wheel invented three times. The molecular structures of the three carbonic anhydrases. EMBO Reports 1: 16–17.

Lima-de-Faria, A. 1991. A kit to produce an artificial human chromosome. In: Fundamentals of Medical Cell Biology, Volume 1. Evolutionary Biology. JAI Press Inc.: 115–161.

Lima-de-Faria, A. 1997. The atomic basis of biological symmetry and periodicity. BioSystems 43: 115–135.

Lima-de-Faria, A. 1998. The role of homeotic genes and of molecular mimicry in the determination of plant and animal symmetries. In Symmetry in Plants (Edited by R.V. Jean and D. Barabé). World Scientific: XXVII–XXXIII.

Lima-de-Faria, A. 2001. Genetic mechanisms involved in the periodicity of flight. Caryologia 54: 189–208.

Lingner, J. et al. 1994. Telomerase RNAs of different ciliates have a common secondary structure and a permuted template. Genes Dev 8: 1984–1998.

Little, P. 1999. The book of genes. Nature 402: 467–468.

Liu, N. et al. 1998. XRCC2 and XRCC3, new human Rad51-family members, promote chromosome stability and protect against DNA cross-links and other damages. Mol Cell 1: 783–793.

Lodish, H. et al. 1995. Molecular Cell Biology. Scientific American Books, W.H. Freeman and Company, New York.

Lozano, G. and Elledge, S.J. 2000. p53 sends nucleotides to repair DNA. Nature 404: 24–25.

Luger, K. et al. 1997. Crystal structure of the nucleosome core particle at 2.8 Å resolution. Nature 389: 251–260.

MacDonald, M.E. et al. 1993. A novel gene containing a trinucleotide repeat that is expanded and unstable on Huntingtons-disease chromosomes. Cell 72(6): 971–983.

Mahlknecht, U. et al. 2000. Chromosomal organization and localization of the human histone deacetylase 5 gene (HDAC5). Biochimica et Biophysica Acta 1493: 342–348.

Malicki, J. et al. 1990. Mouse hox-2.2 specifies thoracic segmental identity in *Drosophila* embryos and larvae. Cell 63: 961–967.

Mandel, J.-L. 1997. Breaking the rule of three. Nature 386: 767–769.

Maroni, G. 1993. An Atlas of *Drosophila* Genes. Oxford University Press, Oxford.

Martin, A.B. and Schultz, P.G. 1999. Opportunities at the interface of chemistry and biology. TIG Millennium issue 15(12): M24–M28.

Miki, Y. et al. 1994. A strong candidate for the breast and ovarian-cancer susceptibility gene BRCA1. Science 266: 66–71.

Moazed, D. 2001. Common themes in mechanisms of gene silencing. Molecular Cell 8: 489–498.

Moritz, M. et al. 1995. Microtubule nucleation by gamma-tubulin-containing rings in the centrosome. Nature 378: 638–640.

Morral, N. et al. 1994. The origin of the major cystic-fibrosis mutation (delta-F508) in European populations. Nature Genetics 7(2): 169–175.

Mullins, M.C. et al. 1994. Large-scale mutagenesis in the zebrafish – In search of genes-controlling development in a vertebrate. Current Biology 4: 189–202.

Munoz, M. de L. et al. 2001. Entamoeba histolytica trophozoites activated by collagen type I and Ca^{2+} have a structured cytoskeleton during collagenase secretion. Cell Motility and the Cytoskeleton 50: 45–54.

Murray, A. and Hunt, T. 1993. The Cell Cycle: An Introduction. W.H. Freeman and Company.

Nealson, K.H. and Conrad, P.G. 1999. Life: past, present and future. Phil Trans R Soc Lond B 354: 1923–1939.

Noller, H.F. 1991. Ribosomal RNA and translation. Annu Rev Biochem 60: 191–227.

O'Brien, S.J. et al. 1997. Comparative genomics: lessons from cats. TIG 13(10): 393–399.

Oliver, S.G. et al. 1992. The complete DNA sequence of yeast chromosome III. Nature 357: 38–46.

Orr, W.C. and Sohal, R.S. 1994. Extension of life-span by overexpression of superoxide-dismutase and catalase in *Drosophila melanogaster*. Science 263: 1128–1130.

Oudet, P. et al. 1975. Electron microscopic and biochemical evidence that chromatin structure is a repeating unit. Cell 4: 281–300.

Pasyukova, E.G. et al. 1998. The relationship between the rate of transposition and transposable element copy number for copia and Doc retrotransposons of *Drosophila melanogaster*. Genet Res, Camb 72: 1–11.

Patel, S. and Latterich, M. 1998. The AAA team: related ATPases with diverse functions. Trends in Cell Biology 8: 65–71.

Pennisi, E. 1996. Linker histones, DNA's protein custodians, gain new respect. Science 274: 503–504.

Pereira, L. et al. 1993. Genomic organization of the sequence coding for fibrillin, the defective gene-product in Marfan-syndrome. Human Molecular Genetics 2(7): 961–968.

Pikaard, C.S. 2000. The epigenetics of nucleolar dominance. TIG 16(11): 495–500.

Pluta, A.F. et al. 1995. The centromere: Hub of chromosomal activities. Science 270: 1591–1594.

Quinn, H.R. and Witherell, M.S. 1998. The asymmetry between matter and antimatter. Scientific American October 1998: 50–55.

Rivera-Pomar, R. et al. 1996. RNA binding and translational suppression by bicoid. Nature 379: 746–749.

Roeder, G.S. and Bailis, J.M. 2000. The pachytene checkpoint. TIG 16(9): 395–403.

Ruf, S. et al. 2001. Stable genetic transformation of tomato plastids and expression of a foreign protein in fruit. Nature Biotechnology 19: 870–875.

Safiejko-Mroczka, B. and Bell, P.B., Jr. 2001. Reorganization of the actin cytoskeleton in the protruding lamellae of human fibroblasts. Cell Motility and the Cytoskeleton 50: 13–32.

SanMiguel, P. et al. 1996. Nested retrotransposons in the intergenic regions of the maize genome. Science 274: 765–768.

Schellenberg, G.D. 1995. Genetic dissection of Alzheimer-disease, a heterogeneous disorder. P.N.A.S. 92(19): 8552–8559.

Scherthan, H. 1990. Localization of the repetitive telomeric sequence $(TTAGGG)_n$ in two muntjac species and implications for their karyotypic evolution. Cytogenet Cell Genet 53: 115–117.

Scherthan, H. et al. 1998. Aspects of three-dimensional chromosome reorganization during the onset of human male meiotic prophase. Journal of Cell Science 111: 2337–2351.

Schimmel, P. and Kelley, S.O. 2000. Exiting an RNA world. Nature Structural Biology 7: 5–7.

Schmidt, T. and Heslop-Harrison, J.S. 1998. Genomes,

genes and junk: the large-scale organization of plant chromosomes. Trends in Plant Science 3: 195–199.

Schuler, G.D. et al. 1996. A gene map of the human genome. Science 274: 540–546.

Schwabe, J.W.R. and Travers, A.A. 1993. What is evolution playing at? Curr Bio 3(9): 628–630.

Scott, M.P. 1992. Vertebrate homeobox gene nomenclature. Cell 71: 551–553.

Sharpe, M.E. and Errington, J. 1999. Upheaval in the bacterial nucleoid. An active chromosome segregation mechanism. TIG 15(2): 70–74.

Sherrington, R. et al. 1995. Cloning of a gene bearing missense mutations in early-onset familial Alzheimers-disease. Nature 375: 754–760.

Shore, D. 1997. Telomere length regulation: getting the measure of chromosome ends. Biol Chem 378: 591–597.

Sjöstrand, F.S. 1999. Molecular pathology of Luft disease and structure and function of mitochondria. J Submicrosc Cytol Pathol 31(1): 41–50.

Slijepcevic, P. 1998. Telomere length and telomere-centromere relationships? Mutation Research 404: 215–220.

Slijepcevic, P. et al. 1997. Instability of CHO chromosomes containing interstitial telomeric sequences originating from Chinese hamster chromosome 10. Cytogenet. Cell Genet 76: 58–60.

Slijepcevic, P. et al. 1997. Telomere length, chromatin structure and chromosome fusigenic potential. Chromosoma 106: 413–421.

Somerville, C. and Dangl, J. 2000. Plant biology in 2010. Science 290: 2077–2078.

Sorkin, A. 2000. The endocytosis machinery. Journal of Cell Science 113(24): 4375-4376.

Stormo, G.D. and Fields, D.S. 1998. Specificity, free energy and information content in protein-DNA interactions. Trends Biochem Sci 23: 109–113.

Story, R.M. et al. 1993. Structural relationship of bacterial RecA proteins to recombination proteins from bacteriophage-T4 and yeast. Science 259: 1892–1896.

Strachan, T. and Read, A.P. 1996. Human Molecular Genetics. BIOS Scientific Publishers, Oxford.

Sugiura, M. 1992. The chloroplast genome. Plant Mol Biol 19: 149–168.

Sullivan, B.A. and Willard, H. 1998. Stable dicentric X chromosomes with two functional centromeres. Nature Genetics 20: 227–228.

Tanaka, H. et al. 2000. A ribonucleotide reductase gene involved in a p53-dependent cell-cycle checkpoint for DNA damage. Nature 404: 42–49.

Tarn, W.-Y. and Steitz, J.A. 1997. Pre-mRNA splicing: the discovery of a new spliceosome doubles the challenge. Trends Biochem. Sci 22: 132–137.

Tartaglia, L.A. et al. 1995. Identification and expression cloning of a leptin receptor, OB-R. Cell 83(7): 1263–1271.

The Arabidopsis Genome Initiative. 2000. Analysis of the genome sequence of the flowering plant *Arabidopsis thaliana*. Nature 408: 796–815.

Tully, T. et al. 1994. Genetic dissection of consolidated memory in *Drosophila*. Cell 79(1): 35–47.

Vafa, O. and Sullivan, K.F. 1997. Chromatin containing CENP-A and alpha-satellite DNA is a major component of the inner kinetochore plate. Curr Biol 7: 897–900.

van der Ploeg, L.H.T. 1990. Antigenic variation in African trypanosomes: Genetic recombination and transcriptional control of VSG genes. In: Gene Rearrangement (Eds. B.D. Hames and D.M. Glover). Oxford University Press, New York: 51–97.

Varmus, H. and Weinberg, R.A. 1993. Genes and the Biology of Cancer. Scientific American Library, New York.

Venter, J.C. et al. 2001. The sequence of the human genome. Science 291: 1304–1351.

Verkerk, A.J.M.H. et al. 1991. Identification of a gene (FMR-1) containing a CGG repeat coincident with a breakpoint cluster region exhibiting length variation in fragile-X syndrome. Cell 65: 905–914.

Vigers, G.P.A. and Lohka, M.J. 1991. A distinct vesicle population targets membranes and pore complexes to the nuclear envelope in *Xenopus* eggs. Journal Cell Biology 112: 545–556.

Villena, F.P.M. de et al. 2000. Natural selection and the function of genome imprinting: beyond the silenced minority. TIG 16: 573–578.

Wells, R.A. et al. 1990. Telomere-related sequences at interstitial sites in the human genome. Genomics 8: 699–704.

Wheeler, R.T. and Shapiro, L. 1997. Bacterial chromosome segregation: is there a mitotic apparatus? Cell 88: 577–579.

Wichman, H.A. et al. 1991. Genomic distribution of heterochromatic sequences in equids: Implications to rapid chromosomal evolution. Journal of Heredity 82: 369–377.

Wichman, H.A. et al. 1992. Transposable elements and the evolution of genome organization in mammals. Genetica 86: 287–293.

Wilmut, I. et al. 1997. Viable offspring derived from fetal and adult mammalian cells. Nature 385: 810–812.

Zhang, Y.Y. et al. 1994. Positional cloning of the mouse obese gene and its human homolog. Nature 372: 425–432.

Sources of illustrations

The permission for the reproduction of the material listed below is gratefully acknowledged.
When a fee was requested, payment was made by check.
The *page number* refers: either to the page on which the figure appears in the publication or to the first page of the article in which the figure is included.

Front cover	G.D. Schuler et al. 1996. Science 274: 540–546 (Fig. of chromosome 1).
Front cover	D.M. Mottier 1903. Bot. Gaz. 35: 250–280 (Fig. First meiotic division in *Lilium*).
Front cover	P. Little 1999. Nature 402: 467 (Fig. 1).
Back cover	A.K. Kleinschmidt et al. 1962. Biochim. Biophys. Acta 61: 857 (Fig. Electron micrograph of a T2 bacteriophage).
Back cover	A. Lima-de-Faria et al. 1986. BioSystems 19: 185–212 (Fig. 11).
Page 2	I. Dunham 2000. Trends in Genetics 16(10): 458 (Fig. 1).
Page 5	T. Schwann 1839. In C. Singer 1934, Histoire de la Biologie: 365 (Fig. 132) Payot, Paris.
Page 9	W. Waldeyer 1888. Archiv Mikroskop. Anatomie 32: 18 (Fig. 7).
Page 13	E.M. Wallace 1940. In A.H. Sturtevant and G.W. Beadle 1940, An Introduction to Genetics: 64 (Fig. 21) Saunders Co. Philadelphia.
Page 17	O.T. Avery, C.M. Macleod, and M. McCarty 1944. In G.S. Stent 1971, Molecular Genetics: 178 (Fig. 7.3) Freeman, San Francisco.
Page 21	F. Jacob and E.L. Wollman 1961. Sexuality and the Genetics of Bacteria: 154 (Fig. 25) Academic Press, New York.
Page 27	A. Garcia-Bellido et al. 1979. Scientific American 241: 90 (Fig. Compartments in the development of *D. melanogaster*).
Page 31	T.R. Cech and B.L. Bass 1986. Annu. Rev. Biochem. 55: 615 (Fig. 6 Structural model of a group I intron).
Page 35	P. Bork and R. Copley 2001. Nature 409: 819 (Fig. 1).
Page 39	C. Somerville and J. Dangl 2000. Science 290: 2077 (Table 1).
Page 39	T.A. Brown 1999. Genomes: 135 (Box 6.4) (Based on Strachan and Read 1996) Bios Sci. Publ., Oxford.
Page 39	T. Strachan and A.P. Read 1996. Human Molecular Genetics: 148 (Fig. 7.1 Organization of the human genome) BIOS Scientific Publ., Oxford.
Page 50	L.C. Martin and B.K. Johnson 1949. Practical Microscopy: 92 (Fig. 77) Blackie & Son, London.
Page 50	L.C. Martin and B.K. Johnson 1949. Practical Microscopy: 68 (Fig. 58) Blackie & Son, London.
Page 50	L.C. Martin and B.K. Johnson 1949. Practical Microscopy: 19 (Fig. 21) Blackie & Son, London.
Page 50	M. Langeron 1934. Precis de Microscopie: 134 (Fig. 70) Masson et C., Paris.
Page 50	C.D. Darlington and L.F. La Cour 1960. The Handling of Chromosomes: 27 (Fig. 1) G. Allen and Unwin, London.
Page 50	C.D. Darlington and L.F. La Cour 1960. The Handling of Chromosomes: 172 (Fig. 7) G. Allen and Unwin, London.
Page 50	A. Guilliermond et G. Mangenot 1941. Precis de Biologie Vegetale: 34 (Fig. 16) Masson et C., Paris.
Page 50	G.L. Humason 1972. Animal Tissue Techniques: 57 (Fig. 8) W.H. Freeman, San Francisco.
Page 51	M. Langeron 1934. Precis de Microscopie: 949 (Fig. 323) Masson et C., Paris.
Page 51	G.L. Humason 1972. Animal Tissue Techniques: 432 (Fig. 26) W.H. Freeman, San Francisco.
Page 51	K. Belar 1928. Die Cytologischen Grundlagen der Vererbung (Fig. Spermatocytes of *Stenobothrus*), Berlin.
Page 51	E. Strasburger 1924. Handbook of Practical Botany: 436 (Fig. 153) G. Allen & Unwin, London.
Page 51	M. Demerec and B.P. Kaufmann 1961. Drosophila Guide: 24 (Fig. 11) Carnegie Inst. of Washington, Washington D.C.
Page 51	O.J. Eigsti and P. Dustin, Jr. 1955. Colchicin – In Agriculture, Medicine, Biology and Chemistry (Fig. Pollen mitosis) Iowa Univ. Press, Iowa.
Page 51	A.R. Gopal-Ayengar and E.V. Cowdry 1947. In J. Brachet 1957, Biochemical Cytology: 116 (Fig. 46) Academic Press, New York.
Page 51	J.R. Baker 1950. Cytological Technique: 31 (Fig. 4) Methuen & Co., London.
Page 52	L.C. Martin and B.K. Johnson 1949. Practical Microscopy: 120 (facing) (Fig. 89) Blackie & Son, London.
Page 52	P.M. Glover 1984. Gene Cloning: 133 (Fig. 6.3) Chapman and Hall, London.
Page 52	D.S. Hogness et al. 1975. In W.J. Peacock and R.D. Brock (Eds.) 1975, The Eukaryote Chromosome: 10 (Fig. 4) Australian Nat. Univ. Press, Canberra.
Page 52	D.C. Pease 1960. Histological Techniques for Electron Microscopy: 92 (Fig. 11) Academic Press, New York.
Page 52	A.T. Sumner 1977. In A. De La Chapelle and M. Sorsa (Eds.) 1977, Chromosomes Today Vol. 6: 52 (Fig. 1) Elsevier/North-Holland Biomed. Press, Amsterdam.
Page 52	G.L. Brown et al. 1950. Nature 165: 601 (Fig. Paper chromatogram).
Page 52	H.D. Springall 1956. Chromosomes: 33 (Fig. 4) Uitgevers-Maatschappij, Tjeenk Willink, Zolle, Holland.
Page 53	G.A. Boyd 1955. Autoradiography in Biology and Medicine: 213 (Fig. 74) Acad. Press, New York.
Page 53	J.H. Taylor 1958, 1963. Molecular Genetics 1963, Part I: 74 (Fig. Semiconservative replication of eukaryotic chromosomes) Acad. Press, New York.
Page 53	A. Lima-de-Faria 1959. J. Biophysical and Biochemical Cytology 6: 457–466 (Figs. 1, 4, 7, 10).
Page 53	D.S. Hogness et al. 1975. In W.J. Peacock and A.D. Brock (Eds.) 1975, The Eukaryotic Chromosome: 8 (Fig. 3) Australian Nat. Univ. Press, Canberra.
Page 53	Mintz, B. 1971. In J.C. Daniel (Ed.) 1971, Methods in Mammalian Embryology (Fig. Diagram of procedure for producing allophenic mice) Freeman, San Francisco.
Page 54	L.C. Martin and B.K. Johnson 1949. Practical Microscopy: 108 (Fig. 84) Blackie & Son, London.
Page 54	T. Caspersson 1936. Skand. Arch. Physiol. 73 (Fig. Ultraviolet absorption of DNA and RNA).
Page 54	M. Langeron 1934. Precis de Microscopie: 987 (Fig. 348) Masson & C., Paris.

Page 54	N. Andresen 1942. Compt. Rend. Trav. Lab. Carlsberg, Serie Chim. 24: 140 (Fig. Centrifuged unstained *Amoeba*), Carlsberg Lab., Copenhagen.
Page 54	J.R.B. Hastings 1972. In G.D. Birnie (Ed.) 1972, Subcellular Components: 267 (Fig. 6) Butterworths, London.
Page 54	S. Osawa and A. Sibatani 1967. In L. Grossman and K. Moldave 1967, Methods in Enzymology, Vol. 12A: 683 (Fig. 3) Academ. Press, New York.
Page 54	H. Lodish et al. 1995. Molecular Cell Biology: 165 (Fig. 5.29) Sci. Am. Books, Freeman and Co., New York.
Page 54	R.E. Franklin and R. Gosling 1953. Nature 171: 740 (Fig. X-Ray diffraction photograph of DNA).
Page 55	J. Brachet 1957. Biochemical Cytology: 6 (Fig. 1) Acad. Press, New York.
Page 55	K. Arms and P.S. Camp 1995. Biology: 278 (Fig. 13-D) Saunders Coll. Publ., Philadelphia.
Page 55	D.M. Jr. Fambrough 1969. In A. Lima-de-Faria (Ed.) 1969, Handbook of Molecular Cytology: 453 (Fig. 3) North Holland Publ. Co., Amsterdam.
Page 55	A.A. Sandberg 1980. The Chromosomes in Human Cancer and Leukemia: 36 (Fig. 15A) Elsevier, Amsterdam.
Page 55	R.W. Old and S.B. Primrose 1980. Principles of Gene Manipulation: 6 (Fig. 1.2) Blackwell Sci. Publ., Oxford.
Page 55	T. Friedman 1971. Scientific American 1971 November (Fig. Amniocentesis).
Page 55	K. Arms and P.S. Camp 1995. Biology: 277 (Fig. 13–16) Saunders College Publ., Philadelphia.
Page 56	D.M. Glover 1984. Gene Cloning: 210 (Fig. 8.17) Chapman and Hall, London.
Page 56	T.A. Brown 1999. Genomes: 54 (Fig. 3.13) Bios Scient. Publ., Oxford.
Page 56	K. Arms and P.S. Camp 1995. Biology: 265 (Fig. 13(B,C)) Saunders College Publ., Philadelphia.
Page 56	B. Alberts et al. 1994.. Molecular Biology of the Cell: 317 (Fig. 7-32) Garland Publ., New York.
Page 56	O.J. Miller and A.H. Bakken 1972. Gene Transcription in Reproductive Tissues, 5th Symposium, Symposia in Reproductive Endocrinology: 155 (Fig. An electron micrograph of a transcribing ribosomal gene) Karolinska Inst., Stockholm.
Page 60	R. Hooke 1665. In C. Singer 1934, Histoire de la Biologie: 356 (Fig. 130) Payot, Paris.
Page 60	T. Schwann 1847. In W. Coleman 1979, Biology In the Nineteenth Century: 27 (Fig. 2.2) Cambridge Univ. Press, Cambridge.
Page 60	E.B. Wilson 1925. The Cell in Development and Heredity: 23 (Fig. 6) Macmillan Co., New York.
Page 60	S.W. Hurry 1965. The Microstructure of Cells (Fig. Diagrammatic generalized eukaryotic cell) John Murray, London. In J.D. Bernal 1967, The Origin of Life: 134 (Fig. 7) Weidenfeld and Nicolson, London.
Page 60	D. Fawcett 1971. In G.S. Stent 1971, Molecular Genetics: 41 (Fig. 2.8) W.H. Freeman, San Francisco.
Page 61	C. De Duve 1984. A Guided Tour of the Living Cell, vol. 1: 19 (Fig. Map of the cell) Sci. Am. Library, Sci. Am. Books, New York.
Page 61	A.P. Pugsley 1989. Protein Targeting: 2 (Fig. 1.1) Academic Press, New York.
Page 61	T. Cavalier-Smith 1991. In E.E. Bittar (Ed.) 1991, Evolutionary Biology, vol. 1: 233 (Table 1) JAI Press, Greenwich.
Page 61	A. Drawid et al. 2000. TIG 16(10): 426 (Fig. 1) Elsevier Science, Amsterdam.
Page 61	P. Jones 2001. New Scientist, March 17, 2001: 3 (Fig. 3).
Page 62	K. Arms and P.S. Camp 1995. Biology: 99 (Fig. 5-8) Saunders College Publ., Philadelphia.
Page 62	S. Patel and M. Latterich 1998. Trends in Cell Biology 8 (Febr.): 69 (Fig. 3).
Page 62	A. Sorkin 2000. Journal of Cell Science 113(24): 4375 (Fig. Major proteins involved in regulation of endocytic pathways).
Page 64	E.G. Balbiani 1885. In E.B. Wilson 1925, The Cell in Development and Heredity: 660 (Fig. 316) Macmlllan, New York.
Page 64	M. Verworn 1888. In E.B, Wilson 1925, The Cell in Development and Heredity: 658 (Fig. 315) Macmillan, New York.
Page 64	J. Hämmerling 1953. Intern. Rev. Cytology, vol. 2 (Fig. Diagrammatic representation of the influence of the nucleus) Academic Press, New York.
Page 64	J.B. Gurdon 1974. The Control of Gene Expression in Animal Development: 17 (Fig. 5) Clarendon Press, Oxford.
Page 64	J.B. Gurdon 1974. The Control of Gene Expression in Animal Development: 22 (Fig. 7A) Clarendon Press, Oxford.
Page 64	R. Briggs and T.J. King 1953, J. Exptl. Zoology 122: 485 (Fig. Method of preparing hybrid blastulae).
Page 66	W. Hofmeister 1848. In L.W. Sharp 1943, Fundamentals of Cytology: 4 (Fig. 2) McGraw-Hill Book Co., New York.
Page 66	E. Van Beneden and J. Julin 1884. In E.B. Wilson 1925, The Cell in Development and Heredity: 179 (Fig. 76) Macmillan, New York.
Page 66	W. Fleming 1882. In W. Coleman 1979, Biology.in the Nineteenth Century: 39 (Fig. 3.2) Cambridge Univ. Press, Cambridge.
Page 66	E.B. Wilson 1925. The Cell in Development and Heredity: 118 (Fig. 45) Macmillan, New York.
Page 66	M. Heidenhain 1896. In E.B. Wilson 1925, The Cell in Development and Heredity: 128 (Fig. 52) Macmillan, New York.
Page 66	E.B. Wilson 1925. The Cell in Development and Heredity: 5 (Fig. 2) Macmillan, New York.
Page 66	E.B. Wilson 1925. The Cell in Development and Heredity: 3 (Fig. 1) Macmillan, New York.
Page 69	W. Fleming 1887 and F. Meves 1896. In E.B. Wilson 1925, The Cell in Development and Heredity: 507 (Fig. 242) Macmillan, New York.
Page 69	L. Guignard 1898. In L. Blaringhem 1937, Hérédité, Mutation et Evolution: 11 (Fig. 2) Masson et C., Paris.
Page 69	C.D. Dariington 1937. Recent Advances in Cytology: 88 (Fig. 21A) Churchill, J. & A., London.
Page 69	L.W. Sharp 1934. Introduction to Cytology: 257 (Fig. 150) McGraw-Hill Book, New York.
Page 70	M.J. Moses 1968. In M.J.D. White 1973, Animal Cytology and Evolution: 156 (Fig. 6.3 Diagram of a frontal section of a synaptonemal complex) Cambridge Univ. Press, Cambridge.
Page 70	M. Westergaard and D. Von Wettstein 1970. C.R. Lab. Carlsberg 37: 239 (Fig. Electron micrograph of the ultrastructure of the synaptonemal complex).
Page 70	O. Hess 1971. In H.-W. Altmann et al. (Edit.) 1971, Handbuch der Allgemeinen Pathologie, vol. 2: 215 (Fig. The Y chromosome of *D. hydei*) Springer-Verlag, Berlin.
Page 70	D.L. Lindsley and E. Lifschytz 1971. In R.A. Beatty and S. Gluecksohn-Waelsch (Edit.) 1971, Proc. Internat. Symp. on the Genetics of the Spermatozoon: 203 (Fig. The male germinal cycle).
Page 70	A. Weismann 1893. In C.D. Darlington 1953, The Facts of Life: 81 (Fig. 4 Weismann's view of recombination among chromosomes) G. Allen and Unwin, London.
Page 70	J.E. Haber 1999. TIBS (July) 24: 271 (Fig. 1 Repair of doublestrand breaks).

Page 72	V. Gregoire 1932. In G. Tischler 1951, Allgemeine Pflanzenkaryologie, vol. II (2,1): 111 (Fig. 33) Berlin Nikolassee, Bomtraeger.
Page 72	F.W. Tinney 1935. In C.D. Darlington 1937, Recent Advances in Cytology: 308 (Fig. 100) J.&A. Churchill, London.
Page 72	D.H. Wenrich 1916. Bull. Mus. Comp. Zool. Harvard College 60: 55 (Fig. Chromosome B from spermatocytes).
Page 72	H.D. Berendes 1965. Chromosoma 17: 35 (Fig. Drawings of region 115–119 . . .).
Page 72	A. Lima-de-Faria et al. 1959. Hereditas 45: 475 (Figs. 26–40).
Page 72	O. Hess 1971. Handbuch der Allgemeine Pathologie 2: 215 (Fig. Model of the organization of the chromomere) Springer-Verlag, Berlin.
Page 78	F. Schrader 1944. Mitosis: 22 (Fig. 5) Columbia Univ. Press, New York.
Page 78	A. Lima-de-Faria 1949. Hereditas 35: 78 and 79 (Figs. 1 and 2).
Page 78	K.R. Lewis and B. John 1963. Chromosome Marker: 34 (Fig. 11) J.&A. Churchill, London.
Page 78	E. Niebuhr and F. Skovby 1977. Hereditas 86: 121 (Fig. The centromere structure after breakage).
Page 78	A.F. Pluta et al. 1995. Science 270: 1592 and 1593 (Figs. 2 and 3).
Page 79	R.S. Lillie 1905,1911. In F. Schrader 1944, Mitosis: 53 (Fig. 12) Columbia Univ. New York.
Page 79	B.R. Brinkley and E. Stubblefield 1970. In Prescott et al. (Eds.) 1970. Advances in Cell Biology, vol. 1: 119 (Fig. Diagrammatic representation of kinetochore) North-Holland Publ., Amsterdam.
Page 79	A. Lima-de-Faria 1955. Hereditas 41: 238 (Figs. 5–8).
Page 79	V.J. Goyanes and J.B. Schvartzman 1981. Chromosoma 83: 93 (Fig. Electron micrograph of whole-mounted chromosomes).
Page 79	J. Carbon 1984. Cell 37: 353 (Fig. 2).
Page 79	K.H.A. Choo 1997. The Centromere: 78 (Fig. 4.1) Oxford Univ. Press, Oxford.
Page 79	The Arabidopsis Genome Initiative 2000. Nature 408: 805 (Fig. 6).
Page 83	J. Gelei 1921. In E.B. Wilson 1925, The Cell In Development and Heredity: 568 (Fig. 279) Macmillan, New York.
Page 83	C.D. Darlington 1936. In C.D. Darlington 1937, Recent Advances In Cytology: 91 (Fig. 22) J.&A. Churchill, London.
Page 83	B. McClintock 1944. In C.P. Swanson 1960, Cytology and Cytogenetics: 161 (Fig. 6-5) Macmillan, London.
Page 83	E. Gilson et al. 1993. Trends in Cell Biology 3: 132 (Fig. Telomeres and heterochromatic domains).
Page 83	M. Demerec and M.E. Hoover 1936. In C.P. Swanson 1960, Cytology and Cytogenetics: 157 (Fig. 6-3) Macmillan, London.
Page 83	A. Lima-de-Faria 1963. Evolution 17: 293 (Fig. 13).
Page 83	T. Mello-Sampayo 1966. Agron. Lusitana 25: 439 (Fig. 1 Schematic comparison of several chromosome 10).
Page 84	M. Rhoades 1952. In J.W. Gowen (Ed.) 1952, Heterosis: 66 (Fig. Anaphase II of meiosis in maize). Iowa State College Press, Ames.
Page 84	M.J.D. White 1954. Animal Cytology and Evolution: 27 (Fig. 2) Cambridge Univ. Press, Cambridge.
Page 84	D. Kipling 1995. The Telomere: 121 (Fig. 6.1) Oxford Univ. Press, Oxford.
Page 84	D. Kipling 1995. The Telomere: 9 (Fig. 1.5) Oxford Univ. Press, Oxford.
Page 84	D. Kipling 1995. The Telomere: 37 (Table 3.1) Oxford Univ. Press, Oxford.
Page 84	C. Lee et al. 1993 and H. Scherthan 1990. In Cytogenet. Cell Genet. 63: 158 (Fig. 2).
Page 87	J. Gelei 1921. In E.B. Wilson 1925, The Cell in Development and Heredity: 910 (Fig. 428) Macmillan, New York.
Page 87	S.L. Frolowa 1935. In C.D. Darlington 1937, Recent Advances in Cytology; 176 (Fig. 59) J.&A. Churchill, London.
Page 87	E. Heitz 1931. In E. Heitz 1956, Chromosomes: 11 (Fig. 5) Uitgevers-Maatschappij Tjeenkwillink-Zwolle, Holland.
Page 87	B. McClintock 1934. In G. Tischler 1951, Allgemeine Pflanzenkaryologie, vol. II: 208 (Fig. 117) Berlin-Nikolassee, Bomtraeger, Berlin.
Page 87	T. Caspersson 1950. In J. Brachet 1957, Biochemical Cytology: 343 (Fig. 141) Academ. Press, New York.
Page 87	W.S. Vincent 1957. In J. Brachet 1957, BioChemical Cytology: 132 (Fig. 59) Academ. Press, New York.
Page 88	R.P. Perry 1969. In A. Lima-de-Faria (Ed.) 1969, Handbook of Molecular Cytology: 631 (Fig. 1) North-Holland Publ., Amsterdam.
Page 88	A. Lima-de-Faria 1973. Nature New Biology 241: 136 (Fig. 2).
Page 88	H.F. Noller 1991. Ann. Rev. Biochem. 60: 191 (Fig. Binding sites of RNA on a ribosome).
Page 88	J.C.H. de Man and N.J.A. Noorduyn 1969. In A. Lima-de-Faria (Ed.) 1969, Handbook of Molecular Cytology: 1081 (Table 1) North Holland Publ., Amsterdam.
Page 88	C.S. Pikaard 2000. TIG 16(11): 496 (Fig. 2).
Page 92	M.J. Schleiden 1838. In C. Singer 1934, Histoire de la Biologie: 383 (Fig. 135) Payot, Paris.
Page 92	O. Bütschli 1878. In C. Singer 1934, Histoire de la Biologie: 377 (Fig. 134) Payot, Paris.
Page 92	M. Heidenhain 1897. In L.W. Sharp 1934, Introduction to Cytology: 3 (Fig. 1) McGraw-Hill, New York.
Page 92	H. Curtis 1983. Biology: 107 (Fig. 5.12) Worth Publishers, Inc., New York.
Page 92	J.R. Baker 1950. Cytological Technique: 5 (Fig. 3) Methuen, London.
Page 92	E. Lazarides 1994. In B. Lewin 1994, Genes V: 41 (Fig. 2. 12 Diagram, composition of actin microfilaments) Oxford Univ. Press, Oxford.
Page 92	R.D. Goldman 1995. In H. Lodish et al. 1995, Molecular Cell Biology: 1114 (Fig. 23-58) Sci. Am. Books, Freeman, New York.
Page 94	L.R. Cleveland 1938. In E.J. DuPraw 1970, DNA and Chromosomes: 100 (Fig. 6.9) Holt, Rinehart and Winston, New York.
Page 94	M.J. Moses and J.R. Coleman 1964. In E.J. DuPraw 1970, DNA and Chromosomes: 250 (Fig. 15.1) Holt, Rinehart and Winston, New York.
Page 94	H. Moor 1967. In N. Higashi (Ed.) 1967, The World through the Electron Microscope, vol. III Biology: 88 (Fig. Frozen-etched preparation) Jeol.
Page 94	H. Busch et al. 1964. In M. Locke (Ed.) The Role of Chromosomes in Development: 54 (Table II) Academ. Press, New York.
Page 94	G.P. Vigers and M.J. Lohka 1991. J. Cell Biol. 112: 545 (Fig. Assembly of the nuclear envelope).
Page 94	G.P. Vigers and M.J. Lohka 1991. In H. Lodish et al. 1995, Molecular Cell Biology: 1219 (Fig. 25-16) Sci. Am. Books, Freeman, New York.

Page 94	A. Murray and T. Hunt 1993. The Cell Cycle; An Introduction (Fig. Assembly of nuclear envelope) Freeman, New York.
Page 94	B.J. Stevens and J. Andre 1969. In A. Lima-de-Faria (Ed.) 1969, Handbook of Molecular Cytology: 838 (Fig. 1) North-Holland Publ., Amsterdam.
Page 94	C.W. Akey and M. Radermacher 1993. J. Cell. Biol. 122: 1 (Fig. Structure of the nuclear pore complex).
Page 96	M. Jörgensen 1913. In L.W. Sharp 1934, Introduction to Cytology: 156 (Fig. 86) McGraw-Hill Book, New York.
Page 96	E.B. Wilson 1925. The Cell in Development and Heredity: 399 (Fig. 185) Macmillan, New York.
Page 96	E.J. DuPraw 1970. DNA and Chromosomes: 72 (Fig. 5.1) Holt, Rinehart and Winston, New York.
Page 96	A. Ross 1968. J. Ultrastruct. Res. 23: 537 (Fig. Cross section of a centriole).
Page 96	H. Lodish et al. 1995. Molecular Cell Biology: 1082 (Fig. 23-29 (Diagram)) Sci. Am. Books, Freeman, New York.
Page 96	H. Lodish et al. 1995. Molecular Cell Biology: 1096 (Fig. 23-41 (Diagram)) Sci. Am. Books, Freeman, New York.
Page 96	J.S. Doxsey et al. 1994. Cell 76: 639 (Fig. Microtubule organizing center).
Page 96	J.S. Doxsey et al. 1994. In H. Lodish et al. 1995, Molecular Cell Biology: 1058 (Fig. 23.5) Sci. Am. Books, Freeman, New York.
Page 98	P. Favard 1969. In A. Lima-de-Faria (Ed.) 1969, Handbook of Molecular Cytology: 1147 (Fig. 10) North Holland Publ., Amsterdam.
Page 98	P.J. Goldblatt 1969. In A. Lima-de-Faria (Ed.) 1969, Handbook of Molecular Cytology: 1103 (Fig. 1) North-Holland Publ., Amsterdam.
Page 98	P.J. Goldblatt 1969. In A. Lima-de-Faria (Ed.) 1969, Handbook of Molecular Cytology: 1107 (Fig. 5) North-Holland Publ., Amsterdam.
Page 98	P. Favard 1969. In A. Lima-de-Faria (Ed.) 1969, Handbook of Molecular Cytology: 1132 (Fig. 1) North-Holland Publ., Amsterdam.
Page 98	P. Favard 1969. In A. Lima-de-Faria (Ed.) 1969, Handbook of Molecular Cytology: 1149 (Fig. 11) North-Holland Publ., Amsterdam.
Page 98	J.D. Jamieson and G.E. Palade 1967. J. Cell Biology 34: 577 (Fig. Localization of radioactivity).
Page 100	G.H.A. Clowes 1916. In L.W. Sharp 1934. Introduction to Cytology: 36 (Fig. 17) McGraw-Hill, New York.
Page 100	J.R. Danielli and H.A. Davson 1935. J. Cell Physiology 5: 495 (Fig. Membrane model).
Page 100	J.R. Danielli 1954. Colston Papers 7: 1 (Fig. Diagram of cell membrane).
Page 100	G. Guidotti 1972. Ann. Rev. Biochem. 41: 731 (Table – Chemical composition of some purified membranes).
Page 100	W. Stockem and K.E. Wohlfarth-Botterman 1969. In A. Lima-de-Faria (Ed.) 1969, Handbook of Molecular Cytology: 1392 (Fig. 7) North-Holland Publ., Amsterdam.
Page 100	H. Lodish et al. 1995. Molecular Cell Biology: 596 (Fig. 14.1) Sci. Am. Books, Freeman, New York.
Page 101	O.L. Sponsler 1929. Plant Physiology 4: 329 (Fig. Cell wall).
Page 101	W. Seifriz 1929. Amer. Naturalist 63: 410 (Fig. Cell wall structure).
Page 101	A. Frey-Wyssling 1939. Science Progress 134: 249 (Fig. Scheme of submicroscopic structure of a cellulose fiber).
Page 101	P.A. Roelofsen 1965. In R.D. Preston (Ed.) 1965, Advances in Botanical Research, vol. 2: 69 (Fig, Successive stages of the growth of a young wall) Academ. Press, New York.
Page 101	E. Frei and R.D. Preston 1961. Proc. Royal Soc. B 154: 70 (Fig. Electron micrograph of cell wall).
Page 101	P. Albersheim 1980. In N.E. Tolbert (Ed.) 1980, Biochemistry of Plants, vol. 1: 91 (Fig. Structure of cell wall) Academic Press, New York.
Page 106	R.C. Hart 1955. Proc. Natl. Acad. Sci. U.S. 41: 261 (Fig. Electron micrograph of tobacco mosaic virus).
Page 106	E.A. Jr. Evans 1956. Federation Proc. 15: 827 (Fig. Schematic representation of phage particle).
Page 106	J.H. Subak-Sharpe 1969. In A. Lima-de-Faria (Ed.) 1969, Handbook of Molecular Cytology: 83 (Table 1) North-Holland Publ., Amsterdam.
Page 106	A. Klug and D.L.D. Caspar 1960. Advan. Virus Res. 7: 225 (Fig. Diagram showing the helical arrangement of RNA).
Page 106	J.D. Watson 1970. Molecular Biology of the Gene: 489 (Fig. 15-16) W.A. Benjamin, New York.
Page 106	A.K. Kleinschmidt et al. 1962. Biochim. Biophys. Acta 61 : 857 (Fig. Electron micrograph of a T2 bacteriophage).
Page 107	W.B. Wood and R.S. Edgar 1967. Scientific American July 1967 and W.B. Wood 1973. In F.H. Ruddle (Ed.) 1973, Genetic Mechanisms of Development: 29 (Fig. Morphogenetic pathway of T 4 phage) Academic Press, New York.
Page 107	R.W. Old and S.B. Primrose 1980. Principles of Gene Manipulation: 91 (Fig. 8.2) Blackwell Scient. Publ., Oxford.
Page 107	R.W. Old and S.B. Primrose 1980. Principles of Gene Manipulation: 47 (Fig. 4.1) Blackwell Scient. Publ., Oxford.
Page 107	R. Williams, 1988. In L. Stryer 1988, Biochemistry: 724 (Fig. 30-2) Freeman, New York.
Page 107	A. Klug 1988. In L Stryer 1988, Biochemistry: 725 (Fig. 30-5) Freeman, New York.
Page 107	R.W. Hendrix 1999. Current Biology 9 (No. 24): R 916 (Fig. 3).
Page 110	J. Lederberg and E.L. Tatum 1946. 11th Cold Spring Harbor Symposium Quantitative Biology 1946. In G.S. Stent 1971. Molecular Genetics: 253 (Fig. 10.2) Freeman, San Francisco.
Page 110	J. Cairns 1963. Cold Spring Harbor Symp. Quant. Biol., vol. 28: 43 (Fig. Autoradiograph of the chromosome of *E. coli*) Cold Spring Harbor Laboratory, Cold Spring Harbor, New York.
Page 110	M. Westergaard 1964. C.R. Trav. Lab. Carlsberg 34: 359 (Table – Cytogenetic mechanisms).
Page 110	A.L Taylor and M.S. Thoman 1964. Genetics 50: 667 (Fig. A simplified genetic map of *E. coli*).
Page 110	L.A. MacHattie et al. 1965. J. Mol. Biol. 11: 648 (Fig. Electron micrograph of the bacterium *H. influenza*).
Page 111	L.E. Hood et al. 1975.. Molecular Biology of Eucaryotic Cells, vol. 1: 1 (Fig. 1.1) Benjamin.
Page 111	R.C. Dickson et al. 1975. Science 187: 32 (Figs. 5 and 6 Nucleotide sequence of *Lac* operon).
Page 111	G.S. Stent 1971. Molecular Genetics: 243 (Fig. 9-16) Freeman, San Francisco.
Page 111	R.T. Wheeler and L Shapiro 1997. Cell 88: 577 (Fig. 1).
Page 111	A. Worcel and E. Burgi 1972. J. Mol. Biol. 71: 127 (Fig. Coiling and uncoiling of the *E. coli* chromosome).
Page 114	W. Robyns 1924. La Cellule 34: 367 (Fig. Development of achromatic figure).
Page 114	A. Guilliermond and G. Mangenot 1941. Biologie Vegetale: 96 (Fig. 43) Masson et Cie, Paris.
Page 114	K. Porter 1995. In K. Arms and P.S. Camp 1995, Biology: 107 (Fig. 5-17 (a)) Saunders College, Harcourt Brace College, Philadelphia.

Page 114	I.B. Dawid and D.R. Wolstenholme 1967. J. Mol. Biol. 28: 233 (Fig. Molecules of native DNA from oocyte mitochondria).
Page 114	M. Rabinowitz et al. 1969. In H. Swift and D.R. Wolstenholme 1969. In A. Lima-de-Faria (Ed.) 1969, Handbook of Molecular Cytology: 983 (Fig. 4) North-Holland Publ., Amsterdam.
Page 114	P. Borst and A.M. Kroon 1969. In A.M. Kroon 1969. In A. Lima-de-Faria (Ed.) 1969, Handbook of Molecular Cytology: 951 (Fig. 2) North-Holland Publ., Amsterdam.
Page 114	C.W. Birky, Jr. 1976. BioScience 26: 26 (Fig. Sites of transcription and translation).
Page 115	D.L. Fouts et al. 1975. J. Cell Biol. 67: 378 (Fig. Electron micrograph of kinetoplast DNA).
Page 115	P. Borst and LA. Grivell1981 and others. In H. Lodish et al. 1995, Molecular Cell Biology: 814 (Fig. 19-4 The organization of human and yeast mtDNA) Sci. Am. Books, Freeman, New York.
Page 115	V.K. Ekenrode and C.S. Levings 1986. In Vitro Cell Dev. Biol. 22: 169 (Table Mitochondrial genes).
Page 115	C.S. Levings and G.G. Brown 1989. Cell 56: 171 (Table Mitochondrial genes).
Page 115	S. Anderson et al. 1981 and others. Nature 290: 457 (Table Alterations of the standard genetic code).
Page 115	H. Lodish et al. 1995. Molecular Cell Biology: 175 (Fig. 5-43) Sci. Am. Books, Freeman, New York.
Page 115	F.S. Sjöstrand 1999. J. Submicroscopic Cyt. Pathol. 31: 43 (Fig. 2).
Page 118	A.. Guilliermond 1912. Archives de I'Anatomie Microscopique, Paris XIV (Fig. Origin of mitochondria and chloroplasts).
Page 118	E. Steinman and F. Sjöstrand 1955. Exptl. Cell Research 8: 15 (Fig. Schematic representation of chloroplast).
Page 118	R. Sager and M. Ishida 1963. Proc. Natl. Acad. Sci. U.S. 50: 725 (Fig. Sedimentation profiles of DNA).
Page 118	L.K. Shumway 1970. In E.J. DuPraw 1970, DNA and Chromosomes: 311 (Fig. 19-3) Holt, Rinehart and Winston, New York.
Page 118	D. Branton and R.B. Park 1967. J. Ultrastruct. Res. 19: 283 (Fig. Interpretation of the surface of the thylakoids).
Page 118	S.D. Kung 1977. Annu. Rev. Plant Physiol. 28: 401 (Table, Physicochemical properties of chloroplasts).
Page 119	C.L.F. Woodcock and H. Fernandez-Moran 1968. J. Molecular Biol. 31 (3): 627 (Fig. DNA in chloroplasts).
Page 119	K. Ohyama et al. 1986. Nature 322: 572 (Table, Genes encoded by chloroplast DNA).
Page 119	V.I. Bruskov and M.S. Odintsova 1968. J. Mol. Biol. 32: 471 (Fig. Chloroplast ribosomes).
Page 119	H. Lodish et al. 1995. Molecular Cell Biology: 175 (Fig. 5-44) Sci. Am. Books, Freeman, New York.
Page 119	B. Lewin 1994. Genes V: 740 (Fig. 25-3) Oxford Univ. Press, Oxford.
Page 129	L.R. Cleveland 1938. Biol. Bull. 74: 41 (Fig. Mitosis in *Barbulanympha*).
Page 129	J.P. Turner 1940. Arch. Protist 93: 255 (Fig. Mitochondria in *T. canalifera*).
Page 129	R.C. Brusca and G.J. Brusca 1990. Invertebrates: 132 (Fig. 4 A *Euglena*) Sinauer Associates, Sunderland.
Page 129	L.R. Cleveland 1949. Trans. Amer. Phil. Soc. 39: 1 (Fig. Prophase chromosomes).
Page 129	K. Belar 1922. Arch. Proto 46: 1 (Fig. Meiosis in *Actinophrys*).
Page 129	L.R. Cleveland 1954. In I.B. Raikov 1982, The Protozoan Nucleus: 195 (Fig. 58 d Diakinesis) Springer-Verlag, Berlin.
Page 130	M.A. Gorovsky 1986. In J.G. Gall (Ed.) 1986, The Molecular Biology of Ciliated Protozoa: 243 and 244 (Table .IV (Page 243) and Fig. 1 (Page 244)) Academ. Press, New York.
Page 130	K. Murti and D.M. Prescott 1986. In L.A. Klobutcher and D.M. Prescott 1986. In J.G. Gall (Ed.) 1986, The Molecular Biology of Ciliated Protozoa (Fig. 9 Chromatin from *Holosticha*) Academic Press, New York.
Page 130	Y. Ohnuki 1968. Chromosoma 25: 402 (Fig. Chromosome No. 2).
Page 130	J. Minassian and L.G.E. Bell 1976. In I.B. Raikov 1982, The Protozoan Nucleus: 65 (Fig. 13 d Ultrastructure of nuclei) Springer-Verlag, Berlin.
Page 130	N.V. Vinnikova 1976. In I.B. Raikov 1982, The Protozoan Nucleus: 191 (Fig. 56 E Ultrastructure of meiosis) Springer-Verlag, Berlin.
Page 130	E.W. Daniels et al. 1969. In I.B. Raikov 1982, The Protozoan Nucleus: 39 (Fig. 3 Nuclear pore dimensions) Springer-Verlag, Berlin.
Page 130	T.K. Golder 1976. in I.B. Raikov 1982, The Protozoan Nucleus: 9 (Fig. 2 Ultrastructure of the nuclear envelope) Springer-Verlag, Berlin.
Page 130	B. Bowers and E.D. Korn 1968. In J.B. Raikov 1982, The Protozoan Nucleus: 79 (Fig. 16 A Ultrastructure of a MTOC) Springer-Verlag, Berlin.
Page 131	G. Moyne et al. 1975. In l.B. Raikov 1982, The Protozoan Nucleus: 57 (Fig. 11 D Nucleoli of protozoans) Springer-Verlag, Berlin.
Page 131	T.R. Cech et al. 1983. In T.R. Cech 1986. In J.G. Gall (Ed.) 1986, The Molecular Biology of Ciliated Protozoa: 216, 215 (Fig. 7 Organization of conserved sequences, Fig. 6 Transesterification mechanism) Academic Press, New York.
Page 131	L.A. Klobutcher and D.M. Prescott 1986. In J.G. Gall (Ed.) 1986, The Molecular Biology of Ciliated Protozoa: 147 (Fig. 10) Acad. Press, New York.
Page 131	D. Ammermann 1986. In L.A. Klobutcher and D.M. Prescott 1986. In J.G. Gall (Ed.) 1986, The Molecular Biology of Ciliated Protozoa: 124 (Fig. 3) Academic Press, New York.
Page 135	V. Gregoire 1932. Bull. Acad. Roy. Belg. Cl. d. Sci, 17: 1435 (Fig. Nucleus of *Impatiens*).
Page 135	A.E. Longley 1938, J. Agr. Res. 56: 177 (Fig. Knobs of maize chromosomes).
Page 135	P.M.B. Walker 1971, Progress Biophys. Mol. Biol. 23: 145 (Fig. Life history of a satellite DNA).
Page 135	L.F. La Cour 1951. Heredity 5: 37 (Fig. *Trillium erythrocarpum*).
Page 135	C.D. Darlington 1941. Annals of Botany 5: 203 (Fig. Distribution of chiasmata).
Page 135	P. Eberle 1956. Chromosoma 8: 285 (Fig. *Aeschynanthus tricolor* n = 16).
Page 136	A. Lima-de-Faria 1954. Chromosoma 6: 359 (Figs. 45–47).
Page 136	A. Klug 1985. Proc. R.W. Welch. Fdn. Conf. Chemistry Res. 39: 133 (Fig. Model of condensed chromatin fiber).
Page 136	A.J.F. Griffiths et al. 1993. An Introduction to Genetic Analysis: 352 (Fig. 9-54) Freeman, New York.
Page 136	J.R. Paulson and U.K. Laemmli 1977. Cell 12: 817 (Fig. Electron micrographs of a histone-depleted metaphase chromosome).
Page 139	E.F. Fritsch, R.M. Lawn and T. Maniatis 1980. Cell 19: 959 (Fig. The human genes Alpha and Beta-globin).
Page 139	L.A. Fothergill-Gilmore 1991. In Fundamentals of Medical Cell Biology vol. 1: 176 (Fig. 5 Comparison of exon positions) JAI Press.

Page 139	B. Alberts et al. 1994, Molecular Biology of the Cell: 1095 (Fig. 21-69) Garland Publ., New York.
Page 139	H. Varmus and R.A. Weinberg 1993. Genes and the Biology of Cancer: 150 (Fig. Genes encoding transcription factors crucial to development) Sci. Am. Library, HPHLP, New York.
Page 139	M.P. Scott 1992. Cell 71: 551 (Fig. The HOM complex).
Page 140	W. Gilbert et al. 1987. In B. Alberts et al. 1994, Molecular Biology of the Cell: 390 (Fig. 8-78 B) Garland Publ., New York.
Page 140	M. Dayhoff 1972. Atlas of Protein Sequence and Structure (Fig. Phylogenetic tree of cytochrome c) Natl. Biomedical Res. Fnd., Washington D.C.
Page 140	M. Dayhoff 1969. Atlas of Protein Sequence and Structure, vol. 4: 1 (Fig. Time scale of the events in the evolution of immunoglobulins) Natl. Biomedical Res. Fnd., Silver Spring, MD.
Page 143	B.P. Chowdhary et al. 1998. Genome Research 8: 580 (Fig. 1 Conservation of whole chromosome synteny).
Page 143	S.J. O'Brien et al. 1997. TIG 13 (10): 395 (Fig. 2 Abridged human gene map).
Page 143	M.R. Culbertson 1999. TIG 15 (2): 75 (Fig. 1 Degradation of mRNAs).
Page 143	H. Tanaka et al. 2000. Nature 404: 42 (Fig. p53R2 gene).
Page 143	G. Lozano and S.J. Elledge 2000. Nature 404: 24 (Fig. 1 Differential composition and .localization of RNR).
Page 147	P. Michailova 1976. Caryologia 29(3): 2 (Fig. 1 Salivary chromosome map).
Page 147	A. Lima-de-Faria and H. Jaworska 1964. Hereditas 52: 119 (Figs. 3 and 4 Haploid cells).
Page 147	A. Ferreira et al. 1984. Rev. Brasil. Genet. 8 (2): 233 (Fig. 1 Male meiosis).
Page 147	W.J. Crosland and R.H. Crozier 1986. Science 231: 1278 (Fig. 1).
Page 147	A. Liljas and M. Laurberg 2000. EMBO Reports 1 (1): 16 (Fig. 1).
Page 148	D. Whiteley and M. Nixon 1972. The Oxford Book of Vertebrates: 149 (Figs. 1 and2 Muntjac) Oxford Univ. Press, Oxford.
Page 148	R.E. Cleland 1936. Bot. Rev. 2: 316 (Fig. Chromosome configurations in *Oenothera*).
Page 148	D.H. Wurster and K. Benirschke 1970. Science 168: 1364 (Fig. The mitotic chromosomes).
Page 148	C.C. Chinnappa and R. Victor 1979. Chromosoma 71: 227 (Fig. Sequence of chromosomes in the ring).
Page 148	T.C. Hsu and K. Benirschke 1971. An Atlas of Mammalian Chromosomes, vol. 2: 88 Folio (Fig. The karyotype of *Muntiacus reevesi*) Springer Verlag, Heidelberg.
Page 148	A. Lima-de-Faria et al. 1986. BioSystems 19: 201 (Fig. 11).
Page 152	C.D. Darlington 1937. Recent Advances in Cytology: 59 (Figs. 14 and 15) J.&A. Churchill, London.
Page 152	C.D. Darlington 1932. Chromosomes and Plant Breeding (Fig. Diagram showing what happens to two mating chromosomes) Macmillan, London.
Page 152	M.J.D. White 1973. Animal Cytology and Evolution: 182 (Fig. 6.17 Belling's copy-choice hypothesis) Cambridge Univ. Press, Cambridge.
Page 152	C.B. Bridges 1936. Science 83: 210 (Fig. The bar region).
Page 152	L.W. Sharp 1934. Introduction to Cytology: 303 (Fig. 174) McGraw-Hill Book, New York.
Page 153	T. Dobzhansky 1941. Genetics and the Origin of Species: 126 (Fig. 10) Columbia Univ. Press, New York.
Page 153	K. Patau, 1935. Naturwissenschaften 23: 537 (Fig. Formation of a loop by pairing).
Page 153	B. McClintock 1938. Genetics 23: 315 (Fig. Increase and decrease in size of a ring-chromosome).
Page 153	J. Wang-Peng 1979. In T.C. Hsu 1979, Human and Mammalian Cytogenetics: 145 (Fig. 22.1) Springer-Verlag, Heidelberg.
Page 153	M.E. Drets and H.N. Seuanez 1974. In E. Coutinho and F. Fuchs (Eds.) 1974, Physiology and Genetics and Reproduction, vol. A: 29 (Fig. Diagrammatic representation of the relative size of the heterochromatic segment). Plenum Press, New York.
Page 153	P. San Miguel et al. 1996. Science 274: 765 (Fig. 1).
Page 153	S.N. Cohen and J.A. Shapiro 1980. Scient. Am. 242: 36 (Fig. The duplication that attends the transposon insertion).
Page 156	C.D. Darlington 1935. Proc. Roy. Soc. B 118: 33 (Fig. The uncoiling of relic spirals).
Page 156	C.B. Bridges 1935. J. Heredity 26: 60 (Fig. The chromosome 4 of *D. melanogaster*).
Page 156	S.W. Brown 1949. Genetics 34: 437 (Fig. Diagram to illustrate the changes in the nucleolar chromosome).
Page 156	J.G. Gall 1973. In B.A. Hamkalo and J. Papaconstantinou (Eds.) 1973, Molecular Cytogenetics: 59 (Fig. A comparison of a mitotic with a polytene chromosome of *Drosophila*) Plenum, New York.
Page 156	H. Sass 1980. J. Cell Science 45: 269 (Fig. A schematic representation of the hierarchy of fibrillar organization).
Page 156	J.S. Kaye 1969. In A. Lima-de-Faria (Ed.) 1969, Handbook of Molecular Cytology: 375 (Fig. 7) North-Holland, Amsterdam.
Page 158	J. Rückert 1891. Anatomischer Anzeiger, vol. 1891 (Fig. Chromosomes of the germinal vesicle) Jena.
Page 158	G. Hasitschka 1956. In E. Tschermak-Woess 1971, Handbuch der Allgemeine Pathologie, vol. 2 (2): 569 (Fig. *Papaver rhoeas*, haploid antipode).
Page 158	C. Pavan 1958. Proc. X International Congress of Genetics, vol. 1: 321 (Fig. Chromosomes of *R. angelae*) McGill Univ., Montreal.
Page 158	W. Nagl 1974. Z. Pflanzenphysiol. 73: 1 (Fig. Three suspensor polytene chromosomes).
Page 158	G.L. Stebbins 1971. Chromosomal Evolution in Higher Plants: 12 (Fig. 1.2) Addison-Wesley, Reading, Ms.
Page 158	N. Virkki 1964. Ann. Acad. Sci. Fenn. A IV 75: 1 (Fig. Giant sex chromosomes).
Page 160	G.D. Karpechenko 1928. Zeitschr. Ind. Abst.. Vererb. 48: 1 (Fig. Pods and somatic chromosomes of *Raphanus*).
Page 160	M. Upcott 1936. J. Genetics 33: 135 (Fig. The leaves, fruits and chromosomes of *Aesculus camea*).
Page 160	A.F. Blakeslee 1937. U.S. Dept. Agr. Yearbook, separate 1605 (Fig. The seed capsules of primary heteroploid mutants in *Datura*).
Page 160	C.A. Ninan 1958. Cytologia 23: 291 (Fig. Meiotic prophase (diakinesis) in a sporocyte of *Ophioglossum*).
Page 160	A.H. Sturtevant and G.W. Beadle 1940. An Introduction to Genetics: 256 (Fig. 99 Triploid intersex) W.B. Saunders, Philadelphia.
Page 160	C.B. Jacobson 1970. In E.J. DuPraw 1970, DNA and Chromosomes: 151 (Fig. 9.17 Human diplochromosomes) Holt, Rinehart and Winston, New York.
Page 164	R.J. Britten and D.E. Kohne 1970. Sci. American 222: 24 (Fig. Relationship between the similarity of repeated-DNA sequences and evolutionary history).

Page 164	R.J. Britten and D.E. Kohne 1970. Sci. American 222: 24 (Fig. Time for reassociation of single-copy DNA).
Page 164	A. Lima-de-Faria et al. 1973. Hereditas 73: 133 (Fig. 10).
Page 164	N. Maeda and O. Smithies 1986. Annu. Rev. Genetics 20: 94 (Fig. 3 The evolution of the globin gene family). In B. Alberts et al. 1994, Molecular Biology of the Cell: 389 (Fig. 8-76) Garland Publishing, New York.
Page 164	L.A. Fothergill-Gilmore 1991. Fundamentals of Medical Cell Biology, vol. 1: 178 (Fig. 6 Alternative splicing) JAI Press.
Page 165	W.-H. Li and D. Graur 1991. Fundamentals of Molecular Evolution (Fig. 3 Schematic structure of a typical eukaryotic protein-coding gene) Sinauer Associates, Sunderland, Ms.
Page 165	H. Lodish et al. 1995. Molecular Cell Biology: 502 (Fig. 12.20) Sci. Am. Books, Freeman, New York.
Page 165	F.P.-M. de Villena et al. 2000. TIG 16 (12): 574 (Table 1 Phylogenetic distribution of parental origin effects).
Page 171	N.K. Koltzoff 1928. In 1939, Les Molecules Hereditaires II, Actualités Scientifiques et Industrielles No. 776: 48, Planch III (Fig. 19 Schema du chromosome d'apres N.K. Koltzoff 1928) Hermann and Co., Editeurs, Paris.
Page 171	E. Heitz 1935. Z. Indukt. Abst. U. Vererbungsl. 70: 402 (Fig. Diagram illustrating the concept of chromosome structure).
Page 171	L. W. Sharp 1943. Fundamentals of Cytology: 85 (Fig. 57) McGraw-Hill Book Co., New York.
Page 171	F. Schrader 1944. Mitosis: 8 (Fig. 3) Columbia Univ. Press, New York.
Page 171	E.D.P. De Robertis, W.W. Nowinski and F.A. Saez 1948. General Cytology: 163 (Fig. 68) W.B. Saunders, Philadelphia.
Page 172	C.D. Darlington and K. Mather 1949. The Elements of Genetics: 148 (Fig. 36) George Allen and Unwin, London.
Page 172	A.E. Mirsky and H. Ris 1947. In H. Busch 1962, An Introduction to the Biochemistry of the Cancer Cell (Fig. Schematic representation of a segment of the chromosome) Academic Press, New York.
Page 172	A.E. Mirsky and H. Ris 1947 and 1950. In H. Busch 1962, An Introduction to the Biochemistry of the Cancer Cell (Fig. Schematic representation of a segment of the chromosome) Academic Press, New York.
Page 172	W.T. Astbury 1945. In Le Gros Clark and P.B. Medawar (Eds.) 1945, Essays in Growth and Form: 249 (Fig. The base-stacking structure of DNA) Oxford Univ. Press, Oxford.
Page 172	C.D. Darlington 1950. In C.H. Waddington 1950, An Introduction to Modern Genetics: 365 (Figs. Plate 5C and Fig. 147) George Allen and Unwin, London.
Page 172	A. Engström and J.B. Finean 1958. Biological Ultrastructure: 280 (Fig. VII.14) Academic Press, New York.
Page 173	D.E. Comings 1972. Adv. Human Genet. 3: 237 (Fig. Single stranded model of chromosome structure).
Page 173	R. Kavenoff and B.H. Zimm 1973. Chromosoma 41: 1 (Fig. The DNA molecule runs the entire length of a chromosome).
Page 173	B. Alberts et al. 1994. Molecular Biology of the Cell: 354 (Fig. 8-30) Garland Publ. New York.
Page 173	T. Schmidt and J.S. Heslop-Harrison 1998. Trends Plant Science 3: 195 (Fig. Large scale organization of plant chromosomes).
Page 176	A. Lima-de-Faria 1952. Chromosoma 5: 1–68 (Plate 1).
Page 176	A. Lima-de-Faria 1954. Chromosoma 6: 330–370 (Figs. 3, 4b, 8b).
Page 176	A. Lima-de-Faria 1983. Molecular Evolution and Organization of the Chromosome. Elsevier, Amsterdam, New York.
Page 181	H. Alfven 1966. Worlds-Antiworlds, Antimatter in Cosmology: 26 (Fig. 5) Freeman, San Francisco.
Page 181	H.R. Quinn and M.S. Witherell 1998. Scientific American October 1998: 52 (Fig. Composite particles).
Page 181	R.T. Sanderson 1967. Inorgartic Chemistry: 14 (Fig. Periodic Table) Reinhold Publ., New York.
Page 181	A. Frey-Wyssling 1945. Ernährung und Stoffwechsel der Pflanzen (Fig. Periodic Table of the elements), Zurich, Switzerland.
Page 181	A. Lima-de-Faria 2001. Caryologia 54 (3): 204 (Fig. 15).

Author index

Adams, R.L.P., 188
Agar, W.E., 186
Akey, C.W., 93, 191
Albersheim, P., 99
Alberts, B., 49, 138, 141, 170, 188, 191
Alfvén, H., 123
Allshire, R., 163, 191
Alwine, J.C., 29
Amaldi, F., 25
Ames, B., 28
Ammermann, D., 128
Amon, J.P., 128
Ananiev, E.V., 175
Anderson, S., 32, 113
Anderson, W.F., 37, 191
André, J., 93
Andresen, N., 48
Anfinsen, C.B., 187
Arabidopsis Genome Initiative, 37, 77, 194
Arber, W., 183, 191
Arms, K., 48, 49, 59, 191
Asimov, I., 187
Astbury, W.T., 22, 23, 169
Avery, A.G., 18
Avery, O.T., 19, 169

Babcock, E.B., 157, 186
Bailis, J.M., 68, 86, 193
Bajer, A.S., 77
Baker, J.R., 47, 91
Baker, S.J., 32, 191
Bakken, A.H., 49
Balbiani, E.G., 63
Baltimore, D., 28, 189
Bargiello, T.A., 32
Barnes, R.D., 145
Barrell, B.G., 30
Bartel, D.P., 191
Bass, B.L., 33
Bateson, W., 14, 185
Bauden, F.C., 19
Bauer, H., 18
Baur, E., 185
Bayliss, W.M., 186
Beatty, B.R., 85
Beatty, R.A., 187
Beadle, G.W., 19, 109, 151, 186
Beatty, B.R., 25
Becker, H.J., 71
Beckwith, J.R., 24, 187
Becquerel, A.H., 12
Beermann, W., 25
Belar, K., 47, 71, 128
Bell, A.C., 191
Bell, L.G.E., 128
Bell, P.B., 91, 193
Belling, J., 15, 133, 151, 186
Bender, W., 33
Bendich, A.J., 117
Benirschke, K., 146, 187
Benne, R., 33
Berendes, H.D., 71
Berg, P., 189

Berk, A., 189
Bernal, J.D., 22
Berthold, G., 185
Bertolotto, C.E.V., 75, 191
Berzelius, J.J., 7, 169
Besse, C., 128
Bhattacharyya, M.K., 34, 191
Bickmore, W., 163, 191
Birky Jr., C.W., 113
Birnstiel, M.L., 25, 162
Bisset, K.A., 187
Blackburn, E.H., 30, 41, 191
Blake, C.C.F., 25
Blakeslee, A.F., 15, 18, 159
Blum, B., 33, 191
Bodmer, W.F., 188
Bohr, N., 22
Boivin, A., 19
Bollag, R.J., 36, 191
Bonner, J., 187
Böök, J.A., 128, 159
Borie, N., 151, 191
Bork, P., 36, 43, 191
Borst, B., 63, 113
Bostock, C.J., 188
Botstein, D., 32
Boveri, T., 14, 71, 185
Bowers, B., 128
Boyd, E., 186
Boyd, G.A., 48
Brachet, A. 185
Brachet, J., 18, 47, 63, 187
Bradbury, E.M., 188
Bradley, R.D., 191
Brand, A.H., 123
Branden, C., 189
Branton, D., 117
Bray, D., 188
Brenner, S., 23
Bridges, C.B., 14, 18, 151, 155
Briggs, R., 25, 63
Brinckley, B.R., 76, 146
Britten, R.J., 23, 24, 161
Broda, P., 188
Brown, D.D., 24, 25, 29, 162
Brown, G.G., 113
Brown, G.L., 48
Brown, T.A., 40, 42, 43, 49, 161, 183, 189, 191
Brown, S.W., 155
Brusca, G.J., 128, 191
Brusca, R.C., 128, 191
Bruskov, V.I., 117
Buchner, P., 185
Bult, C.J., 37, 191
Buongiorno-Nardelli, M., 25
Burdon, R.H., 188
Burgi, E., 109
Busch, H., 93, 187, 188
Bush, H., 170
Butenandt, A., 19
Bütschli, O., 91, 185

Cairns, J., 34, 109

Callan, H.G., 175
Calvin, M., 179
Camara, A., 75
Camp, P.S., 48, 49, 59, 191
Campbell, A.M., 188
Cantor, C.R., 33
Capanna, E., 188
Capecchi, M.R., 33
Caprara, M.G., 42, 191
Carbon, J., 41, 77
Carey, J.C., 189
Carlson, E.A., 187
Carr, N.G., 30, 117
Carter, C.R., 145
Case, M.E., 175
Caspar, D.L.D., 105
Caspersson, T.O., 18, 28, 47, 85, 175, 186
Catcheside, D.G., 187
Caullery, M., 185
Cavalier-Smith, T., 59, 191
Cavalli-Sforza, L.L., 188
Cech, T.R., 33, 128
Cedergren, R., 138, 191
Chalfie, M., 32
Chambon, R., 30
Chargaff, E., 20, 23
Chase, M., 105
Chen, C.W., 109, 191
Cheng, E.H.-Y.A., 63, 137, 191
Chevremont, M., 23
Chiarelli, A.B., 188
Chiba, Y., 23, 117
Chikashige, Y., 82, 191
Child, C.M., 185
Chinnappa, C.C., 146
Chomyn, A., 63, 113
Choo, K.H.A., 75, 77, 189, 191
Chow, K.-C., 183, 191
Chowdhary, B.P., 141, 191
Christiano, A.M., 183, 191
Clarke, L., 77
Claude, A., 18, 113
Clausen, J., 186
Clausen, R.E., 186
Clegg, M.T., 189
Cleland, R.E., 15, 146, 151
Cleveland, L.R., 93, 127, 128
Clowes, G.H.A., 99
Cohen, S.N., 28, 151
Coleman, J.R., 93
Collet, M.S., 28
Collins, F.S., 189
Comings, D.E., 170
Connor, M., 188
Conrad, P.G., 127, 193
Constantini, F., 33
Cook, R.A., 186
Copley, R., 36, 43, 191
Correns, C., 14, 63, 185
Coulondre, C., 30
Coulson, A.R., 29
Coulter, J.M., 185
Covello, P.S., 113
Cowdry, E.V., 47
Crawford, I.P., 137
Creighton, H.B., 18
Crew, F.A.E., 186
Crick, F.H.C., 20, 23, 24, 40, 76, 169, 175
Crosland, M.W.J., 145
Crouse, H.V., 68

Crow, J.F., 68
Crozier, R.H., 145
Cuénot, L., 14
Culbertson, M.R., 141, 191
Cummings, M.R., 189
Curie, M., 12
Curie, P., 12
Curtis, H., 91
Cuvier, G.L.C., 6, 7

Daguerre, L.J.M., 7
Dalton, A.J., 22
Dalton, J., 7
D'Amato, F., 188
Dana, K., 28
Dangl, J., 43, 194
Danielli, J.F., 99, 187
Daniels, E.W., 128
Darlington, C.D., 15, 19, 47, 67, 68, 75, 76, 81, 134, 151, 155, 169, 175, 186, 187
Darnell, J., 189
Darwin, C., 10
Davidson, E.H., 188
Davidson, J.N., 186
Davis, R.W., 28
Davson, H.A., 99
Dawid, I.B., 113, 162
Dayhoff, M.O., 137, 141, 183
Debergh, P., 29
de Bruin, D., 42, 75, 82, 191
de Buffon, G., 7
de Duve, C., 22, 59, 188
Delbrück, M., 19, 22, 109
de Man, J.C.H., 85
Demerec, M., 47, 81, 175
Denhardt, D.T., 188
De Pomerai, D., 188
De Robertis, E.D.P., 169, 170
de Vries, H., 14, 185
Dickson, R.C., 109
Dobzhansky, Th., 151, 186, 187
Doncaster, L., 185
Donelson, J.E., 184
Doty, P., 24
Doxsey, S.J., 95, 192
Drawid, A., 59, 192
Dressler, D., 188
Drets, M.E., 151
Driesch, H., 185
Driever, W., 34
Dubnau, J., 36, 192
Dudits, D., 29
Duggar, B.M., 186
Dujon, B., 30, 36, 192
Dunham, I., 36, 192
Dunn, L.C., 159, 186
DuPraw, E.J., 95, 187
Dustin, P., 47
Dyer, A.F., 48, 67, 188

Eames, A.J., 186
Earnshaw, A., 179
East, E.M., 185
Eberle, P., 134
Eckenrode, V.K., 113
Edenberg, H.J., 170
Edgar, R.S., 105
Egolina, N.A., 28
Eigsti, O.J., 47
Elledge, S.J., 141, 193

Ellis, E.L., 19
Ellis, H.M., 32
Emerson, R.A., 186
Engel, W., 128
Engstrom, A., 169, 187
Ephrussi, B., 19, 109, 187
Erickson, R.L., 28
Errington, J., 109, 194
Evans, E.A., 105
Evans, H.J., 28
Evans, S.K., 184, 192

Fambrough, D.M., 47
Favard, P., 97
Fawcett, D., 59
Felix, M.D., 22
Ferguson, A., 188
Ferguson-Smith, M., 188
Fernandez-Moran, H., 22, 117
Ferreira, A., 145
Feulgen, R., 15, 169
Fields, D.S., 40, 194
Fiers, W., 28
Finean, J.B., 22, 169, 187
Fischer, A., 186
Fisher, R.A., 15, 186
Fizeau, A.H.L., 7
Flavell, R.A., 30
Fleischmann, R.D., 36, 192
Flemming, W., 11, 65, 67, 185
Ford, C.E., 24
Fothergill-Gilmore, L.A., 137, 162, 192
Foucault, J.B.L., 7
Fouts, D.L., 113
Fox, S.W., 179, 187
Fraenkel-Conrat, H., 25
Franklin, N.C., 175
Franklin, R.E., 48
Fraser, C.M., 36, 192
Freeman, F.N., 186
Frei, E., 99
Freifelder, D., 189
Frey-Wyssling, A., 99, 179, 186
Friedberg, E.C., 188
Friedman, T., 49
Fristrom, J.W., 189
Frolowa, S.L., 85
Frönicke, L., 86, 192

Galilei, Galileo, 14
Gall, J.G., 25, 30, 155, 189
Galton, F., 10
Gamow, G., 22
Garcia-Bellido, A., 29, 48
Garnier, C., 97
Gates, R.R., 185
Gaucher, 32
Gehring, W.J., 33, 161
Gelbart, W.M., 189
Gelei, J., 81, 85
Gelehrter, T.D., 189
German, J., 188
Giardina, A., 162
Gilbert, S.F., 189
Gilbert, W., 23, 28, 29, 30, 137
Giles, N.H., 175
Gilson, E., 81
Gineitis, A.A., 81, 192
Gitschier, J., 32
Glover, D.M., 48, 49, 188

Gobineau, J.A., 10
Goldblatt, P.J., 97
Golder, T.K., 128
Goldman, R.D., 91, 192
Goldschmidt, R., 185, 186, 187
Goldstein, J.L., 137, 192
Goldstein, L., 25
Golgi, C., 12, 22
Gopal-Ayengar, A.R., 47
Gordon, J.W., 33
Gorovsky, M.A., 127
Gosling, R., 48
Goyanes, V.J., 76
Graur, D., 162, 192
Gray, M.W., 113
Green, H., 24
Green, M.C., 137
Green, M.M., 187
Greenwald, I.S., 33
Greenwood, N.N., 179
Gregoire, V., 71, 133
Griffiths, A.J.F., 133, 170, 189, 192
Grivell, L.A., 63, 113
Gros, F., 23
Gross, K., 175
Grunberg-Manago, M., 23
Grunstein, M., 28
Gudjonsson, G., 159
Guidotti, G., 99
Guignard, L., 67
Guilliermond, A., 47, 113, 117, 186
Gurdon, J.B., 24, 25, 63, 188
Gurney, M.E., 37, 192
Gurwitsch, A., 185
Guyénot, E., 186
Gvozdev, V.A., 175

Haber, J.E., 67, 192
Haeckel, E.H., 11
Haldane, J.B.S., 15, 81, 186, 187
Halder, G., 36, 192
Hall, B.D., 23, 24
Hall, C.M., 7
Hall, T.H., 33
Hamerton, J.L., 188
Hämmerling, J., 63
Hanmann, J.F., 22
Hanstein, J., 185
Harper, M.E., 33
Harrington, J., 33, 192
Harris, H., 188
Hart, R.C., 105
Hartl, D.L., 189
Hartmann, M., 185
Hartmann-Goldstein, I.J., 175
Hasitschka, G., 157
Hastings, J.R.B., 48
Hatschek, E., 186
Hauschka, T.S., 128
Hayes, W., 187
Hazen, R.M., 179, 192
Hedges, R.W., 30
Hedrick, P.W., 189
Heidenhain, M., 65, 91
Heim, S., 128, 189
Heitz, E., 18, 85, 133, 169, 175
Hendrix, R.W., 105, 192
Henikoff, S., 34, 192
Herr, W., 32
Hershey, A.D., 105, 188

Hertwig, O., 185
Hertwig, P., 186
Herzenberg, L.A., 49
Heslop-Harrison, J.S., 133, 170, 193
Hess, O., 67, 71
Higa, A., 28
Hiesey, W.M., 186
Hoagland, M.B., 23
Hofmeister, W., 65
Hogness, D.S., 28, 48
Holley, R.W., 24
Holzinger, K.J., 186
Hombrecher, G., 33
Hong, F.D., 32
Hood, L.E., 109
Hooke, R., 59
Hoover, M.E., 81
Hopkins, N.H., 189
Horai, S., 37, 192
Horvitz, H.R., 32
Horowitz, M., 32
Hotta, Y., 29
Howard, A., 23, 48
Hozumi, N., 30
Hsu, T.C., 146, 188
Huberman, J.A., 170, 175
Hudson, T.J., 137, 192
Hughes, A., 187
Hughes, J., 36, 192
Hughes, W.L., 23
Hughes-Schrader, S., 75
Human Genome Project, 175
Humason, G.L., 47
Hunt, T., 189, 193
Hurry, S.W., 59
Hutchison, C.A., 30
Huyashi, Y., 63, 113
Hyden, H., 32

Ijdo, J.W., 37, 192
Illmensee, K., 28
Imai, H.T., 145, 192
Ingram, V,M., 187
International Human Genome Sequencing Consortium, 37, 192
Ishida, M., 117
Itakura, K., 28

Jackson, D.A., 28
Jackson, R.C., 145
Jacob, A.E., 30
Jacob, F., 22, 23, 24, 109, 137, 175
Jacobson, C.B., 159
Jamieson, J.D., 97
Janaki-Ammal, E.K., 186
Jaworska, H., 145
Jeffries, A.J., 32
Johannsen, W., 14
Johansson, C., 28
John, B., 76, 187
John, H., 25
Johnson, B.K., 47, 48
Johnson, R.T., 29
Joliot, F., 19
Joliot-Curie, I., 19
Jones, C.W., 29
Jones, D.F., 185
Jones, K.W., 25
Jones, P., 59, 192
Jorde, L.B., 189

Jörgensen, M., 95
Julin, J., 65

Kaiser, A.D., 28
Kan, W.Y., 28
Kandasamy, M.K., 91, 192
Kandel, E.R., 32
Kaplan, H.S., 28
Karpechenko, G.D., 15, 159
Karström, H., 19
Kaufmann, B.P., 47
Kavenoff, R., 29, 170
Kaye, J.S., 155
Keck, D.D., 186
Keller, E.F., 189
Kelley, S.O., 179, 193
Kemp, J.D., 33
Kendrew, J.C., 25
Kerrebrock, A.W., 37, 192
Khorana, H.G., 24, 28
Kihlman, B.A., 188
Kim, N.W., 36, 192
King, R.C., 189
King, T.J., 25, 63
Kipling, D., 81, 82, 128, 189, 192
Kitajima, E., 117
Kleinschmidt, A.K., 105
Klinefelter., 24
Klobutcher, L.A., 128
Klug, A., 105, 133
Klug, W.S., 189
Knight, R.D., 138, 192
Knighton, D.R., 37, 192
Knoll, M., 18
Koch, H.H.R., 10
Kohne, D.E., 24, 161
Kölliker, A., 7
Koltzoff, N.K., 169, 186
Korn, E.D., 128
Kornberg, A., 23, 188
Kornberg, R.B., 30
Kossel, A., 185
Kroon, A.M., 113
Kudo, R.R., 128, 145
Kuehn, M.R., 32
Kuhn, A., 19
Kung, S.D., 117

La Cour, L.F., 47, 134, 186
Laemmli, U.K., 41, 133, 170
Lamarck, J.B.M., 7
Langer, P.R., 33
Langeron, M., 47, 48
Latta, H., 22
Latterich, M., 59, 193
Laurberg, M., 146, 192
Lauretis, J.A., 117
Lavoisier, A., 7
Lawrence, P.A., 34, 189, 192
Lazarides, E., 91, 192
Learned, R.M., 86
Leavitt, R.W., 188
Leber, 32
Lederberg, J., 109
Lee, C., 81, 82, 192
Leedale, G.F., 128
Lein, J., 19
Levan, A., 24, 128
Levings, C.S., 113
Lewin, B., 117, 188, 189, 192

Lewis, E.B., 29, 75
Lewis, J., 188
Lewis, K.R., 76, 187
Lewontin, R.C., 189
Li, W.-H., 162, 192
Lifschytz, E., 67
Liljas, A., 146, 192
Lillie, F.R., 185
Lillie, R.S., 76
Lima-de-Faria, A., 20, 23, 25, 29, 34, 42, 48, 71, 75, 76, 81, 86, 134, 145, 146, 162, 175, 179, 180, 184, 187–189, 193
Lindsley, D.L., 67
Lingner, J., 41, 193
Little, P., 193
Liu, N., 163, 193
Lobban, P., 28
Locke, M., 187
Lodish, H., 48, 76, 77, 82, 95, 113, 117, 162, 189, 193
Loeb, J., 185
Lohka, M.J., 93, 194
Longley, A.E., 71, 134
Lozano, G., 141, 193
Luger, K., 179, 193
Lundblad, V., 184, 192
Luria, S.E., 19, 109, 175
Lwoff, A., 22
Lyon, M.F., 24

MacDaniels, L.H., 186
MacDonald, M.E., 36, 193
Macdonald, P.M., 34
MacHattie, L.A., 109
MacLean, D.N., 188
MacLeod, C.M., 19
Maeda, N., 162
Mahlknecht, U., 40, 193
Makino, S., 187, 188
Malheiros, N., 75
Malicki, J., 34, 193
Malthus, T.R., 10
Mandel, J.-L., 162, 193
Mandel, M., 28
Mangenot, G., 47, 113
Maniatis, T., 29
Manton, I., 186
Margulis, L., 179, 188
Maroni, G., 37, 193
Martin, A.B., 179, 193
Martin, L.C., 47, 48
Mathaei, J.H., 23
Mather, K., 155, 169, 186
Matsudaira, P., 189
Matsuura, H., 186
Matthews, H., 188
Maxam, A.M., 29
McCarty, M., 19
McClintock, B., 18, 19, 20, 30, 81, 85, 151
McClung, C.E., 14, 186
McGinnis, W., 34
McKusick, V.A., 137, 175, 189
McQuillen, K., 23
Meagher, R.B., 91, 192
Mello-Sampayo, T., 81
Mendel, G., 11, 12, 14, 67
Mendeleev, D.I., 11
Merril, C.R., 28
Mertz, J., 28
Merz, T., 188
Meselson, M., 23
Metchnikoff, E., 10

Meves, F., 67
Meyer, A., 185
Michaelis, A., 187
Michailova, P., 145
Miescher, F., 169
Miki, Y., 36, 193
Miller, J.H., 188, 189
Miller, O.L., 25, 49, 85
Minassian, I., 128
Mintz, B., 28, 48, 63
Miramontes, P., 138, 191
Mirsky, A.E., 41, 133, 170
Mitchell, H.K., 19, 187
Mitelman, F., 128, 188, 189
Mlodzik, M., 161
Moazed, D., 137, 193
Molisch, H., 185
Monier, R., 23
Monod, J., 22, 23, 24, 109, 137, 175
Moor, H., 93
Morange, M., 189
Morgan, T.H., 14, 185, 186
Moritz, M., 37, 193
Morral, N., 37, 193
Moses, M.J., 67, 93
Moyne, G., 128
Muirhead, H., 162
Muller, H.J., 14, 15, 81, 169
Mullins, M.C., 36, 193
Munoz, M.L., 91, 193
Murray, A., 189, 193
Murray, A.W., 33, 183
Murray, K.M., 28
Murray, N.E., 28
Murti, K., 127
Myers, A., 113

Nagl, W., 157, 188
Nathans, D., 28
Navashin, M., 75, 76, 169
Nealson, K.H., 127, 193
Newlands, J.A.R., 11
Newman, H.H., 186
Newton, I., 14
Niebuhr, E., 75, 77
Nilsen, T.W., 42, 191
Ninan, C.A., 159
Nirenberg, M.W., 23
Nitsch, C., 29
Nixon, M., 146
Noll, H., 23
Noll, M., 34
Noller, H.F., 85, 193
Nomura, M., 175
Noorduyn, N.J.A., 85
Novikoff, A.B., 113
Nowell, P., 24
Nüsslein-Volhard, C., 29, 33, 34, 36

O'Brien, S.J., 141, 193
Ochoa, S., 23
Odintsova, M.S., 117
Ohno, S., 187, 188
Ohnuki, Y., 128, 155
Ohyama, K., 32, 63, 117
Okamoto, T., 23
Old, R.W., 49, 105, 137, 188
Oliver, S.G., 36, 193
Olson, L., 28
Oparin, A.I., 179

Orgel, L.E., 179, 188
Orr, W.C., 36, 193
Osawa, S., 48
Oudet, P., 179, 193

Pagels, H.R., 180
Painter, T.S., 18
Palade, G.E., 22, 23, 97, 113
Pardue, M.L., 25, 29
Park, R.B., 117
Parks, D.R., 49
Pasteur, L., 10
Pasyukova, E.G., 151, 193
Pätau, K., 151
Patel, S., 59, 193
Pathak, S., 133
Patterson, J.T., 187
Paulson, J.R., 41, 133, 170
Pavan, C., 25, 157
Pease, D.C., 48
Pelc, S.R., 23, 48
Pelling, C., 25
Pennisi, E., 41, 193
Pereira, L., 36, 193
Perkins, F.O., 128
Perkowska, E., 162
Perry, R.P., 85
Perutz, M.F., 25, 187
Petes, T.D., 170, 183
Pigott, G.H., 30, 117
Pikaard, C.S., 85, 193
Pirie, N.W., 19
Plaut, W., 23
Pluta, A.F., 41, 77, 193
Pollister, A.W., 133
Pontecorvo, G., 187
Porter, K.R., 22, 97, 113
Prescott, D.M., 25, 127, 128
Preston, R.D., 99
Primrose, S.B., 49, 105, 137, 188
Pritchard, D.J., 189
Pugsley, A.P., 59, 97, 189

Quinn, H.R., 123, 193

Rabinowitz, M., 113
Radermacher, M., 93, 191
Raff, M., 188
Raff, R.A., 189
Ramanis, Z., 30
Ranvier, L., 11
Rao, P.N., 29
Raspail, F.V., 7
Rassoulzadegan, M., 32
Ray, D.S., 188
Read A.P., 43, 194
Reddy, E.P., 32
Redfield, H., 75
Renner, O., 151, 186
Resnikoff, W.S., 188
Rhoades, M., 82
Rick,C.M., 137, 175
Rieger, R., 187
Ringertz, N.R., 188
Riggs, A.D., 175
Ris, H., 23, 24, 41, 170
Ritossa, F.M., 24, 25, 85, 161, 184
Rivera-Pomar, R., 36, 193
Roberts, H.F., 186
Roberts, J.W., 189

Roberts, K., 188
Roberts, R.B., 23
Roberts, R.J., 30
Robyns, W., 113
Roderick, T.H., 175
Roeder, G.S., 68, 86, 193
Roelofsen, P.A., 99
Roentgen, W.C., 12
Ross, A., 95
Ross, R., 10
Rossenbeck, H., 15, 169
Rosset, R., 23
Rothman, J.E., 99
Roux, W., 11, 185
Rubin, G.M., 34
Rudkin, G.T., 175
Ruf, S., 117, 193
Rükert, J., 157
Ruska, E., 18
Ruska, H., 105
Russel, L.B., 24
Rutten, M.G., 179, 188

Safiejko-Mroczka, B., 91, 193
Sager, R., 30, 117
Saiki, R.K., 33, 49
Saint-Hillaire, E.G., 7
Sandberg, A.A., 48, 128, 188
Sanger, F., 29, 32
SanMiguel, P., 151, 193
Sass, H., 155
Santesson, B., 128, 159
Saunders, G.F., 33
Savage, R.E., 188
Schellenberg, G.D., 36, 193
Scherthan, H., 49, 81, 82, 86, 175, 192, 193
Schimke, R.T., 30
Schimmel, P., 179, 193
Schleiden, M.J., 7, 59, 91
Schlesinger, M., 18
Schmidt, T., 133, 170, 193
Schoenheimer, R., 22
Schopf, J.W., 127
Schrader, F., 76, 169, 186, 187
Schrödinger, E., 19, 22, 186
Schrevel, J., 128
Schuler, G.D., 175, 194
Schultz, J., 75
Schultz, P.G., 179, 193
Schulz-Schaeffer, J., 188
Schvartzman, J.B., 76
Schwabe, J.W.R., 146, 194
Schwann, T., 7, 59
Schwartz, D.C., 33
Schwartz, J.G., 32
Scott, M.P., 33, 138, 194
Sears, E.R., 75
Seifriz, W., 99
Seuanez, H.N., 151, 188
Shapiro, J.A., 151
Shapiro, L., 109, 194
Sharma, A., 187
Sharma, A.K., 187
Sharp, L.W., 67, 151, 169, 170, 186
Sharp, P.A., 30
Sharpe, M.E., 109, 194
Sherrington, R., 36, 194
Shih, M.-C., 32
Shinozake, K., 32
Shore, D., 82, 194

Shumway, L.K., 117
Sibatani, A., 48
Siekevitz, P., 23
Silber, R., 28
Singer, M., 189
Sinnott, E.W., 159, 186
Sirlin, J.L., 188
Sjöstrand, F.S., 22, 97, 113, 117, 194
Skovby, F., 75, 77
Slijepcevic, P., 34, 41, 81, 194
Slizynska, H., 18
Smellie, R.M.S., 188
Smetana, K., 187
Smith-White, S., 145
Smithies, O., 162
Snoad, B., 157
Snyder, L.A., 189
Sohal, R.S., 36, 193
Sommerville, C., 43, 194
Sørensen, T., 159
Sorkin, A., 59, 194
Southern, E.M., 28, 48
Spallanzani, L., 10
Spiegelman, S., 22–24, 85, 109
Spirin, A.S., 189
Springall, H.D., 48
Stahl, F.W., 23
Stanley, W.M., 19, 105
Stansfield, W.D., 189
Stebbins, G.L., 76, 157, 186, 187
Steffensen, D.M., 29
Steinbeck, K.S., 34
Steinmann, E., 22, 117
Steitz, J.A., 189, 194
Steitz, T.A., 37, 42
Stent, G.S., 109, 187
Stern, C., 18, 151, 186, 188
Stevens, B.J., 93
Stockem, W., 99
Stone, W.W., 187
Stormo, G.D., 40, 194
Story, R.M., 37, 194
Strachan, T., 43, 194
Strasburger, E., 11, 47, 185
Struhl, G., 34, 36, 192
Stubbe, H., 180, 186
Stubblefield, E., 76
Sturtevant, A.H., 14, 151, 159, 186, 187
Subak-Sharpe, J.H., 105
Sugiura, M., 117, 194
Sullivan, B.A., 77, 194
Sullivan, K.F., 41, 194
Sulston, J., 32
Sumner, A.T., 48, 188
Sutton, W.S., 14, 67
Suzuki, D.T., 189
Suzuki, L.P., 29
Svedberg, T., 15
Swanson, C.P., 188
Szostak, J.W., 33, 183
Szybalski, W., 24

Takanami, M., 23
Talleyrand, C.M., 6
Tamarin, R., 188
Tanaka, H., 141, 194
Tarn, W.-Y., 42, 194
Tartaglia, L.A., 37, 194
Tartof, K.D., 161, 184
Tatum, E.L., 19, 109

Taylor, A.L., 109
Taylor, G.R., 145
Taylor, J.H., 23, 48, 188
Temin, H.M., 28
Therman, E., 159, 188
Thoman, M.S., 109
Thomas, M., 48
Thompson, D.W., 185
Thomson, J.J., 12
Timofeyeff-Ressovsky, N.V., 19, 186
Tinney, F.W., 71
Tischler, G., 187
Tiselius, A.W.K., 18
Tjio, J.H., 24
Tobler, H., 162
Tonegawa, S., 30
Tooze, J., 189
Travers, A.A., 146, 194
Treviranus, G.R., 7
Tschermak, E., 14
Ts'o, P., 187
Tsui, L.-C., 32
Tully, T., 37, 194
Turner, H.H., 24
Turner, J.P., 128
Turner, M.J., 184
Tzagoloff, A., 113

Unrau, P.J., 191
Upcott, M.B., 159
Ursprung, H., 188

Vafa, O., 41, 194
van Beneden, E., 11, 65
van Iterson, W., 109
van der Ploeg, L.H.T., 75, 194
Varmus, H., 189, 194
Veda, K., 63, 113
Vendrely, C., 19
Vendrely, R., 19
Venter, J.C., 37, 49, 175, 194
Verkerk, A.J.M.H., 36, 194
Verworn, M., 63
Victor, R., 146
Vigers, G.P., 93, 194
Villena, F.P., 163, 194
Vincent, W.S., 85
Vinnikova, N.V., 128
Virkki, N., 157
von Baer, K.E., 11
von Mohl, H., 7
von Wettstein, D., 67

Waddington, C.H., 186, 187
Wagner, R.P., 187
Waldeyer, W., 11, 185
Walker, P.M.B., 133, 161
Wallace, E.M., 15
Wallace, H., 162
Walmsley, R.M., 183
Wang, L.H., 28
Wang-Peng, J., 151
Warburg, O.H., 185
Warmus, H., 138, 194
Watson, J.D., 20, 23, 24, 40, 76, 105, 169, 175, 187–189
Weaver, R.F., 189
Weber, I.T., 37
Webster, N., 123
Weinberg, R.A., 138, 189, 194
Weiner, A.M., 189

Weismann, A., 67, 185
Weiss, M.C., 24
Wells, R.A., 81, 194
Wenrich, D.H., 71
Westergaard, M., 67, 109
Westmoreland, B.C., 24
Wheeler, R.T., 109, 194
White, M.J.D., 82, 151, 175, 186–188
White, R.L., 189
Whitehouse, H.L.K., 175, 188
Whitely, D., 146
Wichman, H.A., 34, 151, 191, 194
Wieschaus, E., 29, 33
Wiesner, J., 185
Wilde, C.D., 162
Wilkins, M.H.F., 22, 23
Willard, H., 77, 194
Williams, R., 105
Wilmut, I., 37, 194
Wilson, E.B., 11, 59, 65, 95, 186
Wimber, D.E., 29
Winkler, H., 186
Witherell, M.S., 123, 193
Wöhler, F., 7
Wohlfarth-Bottermann, K.E., 99
Wolff, S., 187

Wolman, E.L., 23
Wolstenholme, D.R., 113
Wood, W.B., 105
Woodcock C.L.F., 117
Woods, P.S., 23
Worcel, A., 109
Wright, S., 15
Wurster, D.H., 146
Wylie, A.P., 187

Yao, M.-C., 128
Ycas, M., 187
Young, M.W., 32
Young, W.J., 188
Yu, C.W., 48

Zaenen, I., 28
Zakharov, A.F., 28
Zamecnik, P.C., 23
Zech, L., 28
Zernicke, F., 18
Zhang, Y.Y., 36, 194
Zimm, B.H., 29, 170
Zipser, D., 187
Zipursky, S.L., 189

Subject index

5S ribosomal RNA, 85, 137
5S ribosomal RNA genes, 29
7SKRNA, 42
18S genes, 86
18S ribosomal RNA, 85
28S genes, 86
28S ribosomal RNA, 85

abnormal Philadelphia chromosome, 28
accessory chromosomes, 14, 109
accidental rearrangements, 183
Ac-Ds system, 20
Acetabularia, 63
Acetabularia crenulata, 63
Acetabularia mediterranea, 63
aceto-carmine method, 15
Acheta, 71, 155
Acheta domesticus, 162
actin messenger RNAs, 162
actin monomeres, 91
Actinophrys, 128
activators, 123
additional types of histones, 41
adenine-uracil bonding, 40
adenosine deaminase deficiency, 37
adenosine triphosphate (ATP), 113
adenovirus, 30
Aesculus, 159
agarose gel electrophoresis, 33, 48
aging, 36
Alagoasa, 157
algae, 86, 175
allotetraploid hybrid, 15
alphoid DNA, 41
Alu sequences, 161
Alzheimer's disease gene, 36
Amaranthus, 34
Amaryllidaceae, 157
Amblystoma mexicanum, 134
American mink, 141
amniotic fluid, 49
Amoeba, 25
Amoeba proteus, 128, 145
Amphibia, 183
amplification, 30
anabolic processes, 123
animal cell motility, 91
animal paleontology, 6
antennapedia, 138
antero-posterior axis, 34
anthrax bacillus, 10
antibiotics, 34
antibodies, 123
anticodon, 123
antigens, 123
antihemophilic factor, 32
antioncogenes, 123
antiparallel RNA, 123
antiproton, 123
antisense RNA, 123
antithetical molecular events, 123
antithetical properties, 123
antithetic nature of the chromosome, 71

ants, 145
Aplysia, 32
Arabidopsis, 37, 77
Arabidopsis project, 43
archaeon bacteria, 37
artifacts, 47, 91
artificial chromosomes, 183
artificial radioactivity, 19
Ascaris, 82
atomic constraints, 183
atomic mechanisms, 25
atomic self-assembly, 138
atom theory, 7
attached X-chromosomes, 15
attachment to the nuclear envelope, 81
A. tumefaciens, 33
automated PCR, 2
autoradiography, 23, 28
average size of a human gene, 43
axis of the body, 138

Bacillus subtilis, 109
bacteria, 19, 146
bacterial artificial chromosomes (BACs), 2
bacteriology, 10
bacteriophage lambda, 105
bacteriophage phiX174, 29
bacteriophage Qbeta, 24
bacteriophages, 18, 105
balanced lethals, 15
Balsaminaceae, 146
banding, 28
banding procedure, 48
bands, 71
Barbulanympha ufalula, 127
bar gene, 151
base deletions, 183
base sequence of alanine transfer RNA, 24
base substitutions, 183
basic chromosome pattern, 71
beans, 33
Begonia, 29
beta-globin gene, 33
beta tubulin messenger RNAs, 162
bilateral symmetry of the human body, 179
bioengineering, 48
biogenesis of chloroplasts, 117
biological clock, 32
biotinylated DNA, 33
biotinylated probes, 49
birds, 42, 138
bithorax complexes, 29, 138
bladder carcinoma, 32
B lymphocytes, 30
bobbed phenotype, 161
body segmentation, 29, 33
bouquet stage, 77
Brachycome lineariloba, 145
Brassica, 159
breast and ovarian cancer, 36
bridges, 151
Brownian movement, 59
Bryophyta, 86

BUDR labelling technique, 28

cabbage (*Brassica oleracea*), 15
Caenorhabditis elegans, 32, 33
camera lucida, 47
cancer, 36, 141
cancer cells, 24
carbonic anhydrase, 146
cat, 141
catabolic reactions, 123
catalase, 36
catalitic activity, 179
cattle, 141
cause of malaria, 10
C. elegans, 33
C. elegans sequence, 2
cell chemistry, 47
cell differentiation, 63
cell division, 11
cell fractionation techniques, 18, 113
cell-free systems, 23
cell membranes can self-assemble, 99
cell secretion, 91, 97
cellulose fibers, 99
CENP-A, 41
centrioles, 59, 95
centrioles replicate during G, 95
centromere, 20, 41,
centromere-binding factor (CBF), 77
centromere cycle of division, 76
centromere has a strictly terminal position, 82
centromere pattern, 20
centromere-spindle apparatus, 77
centromere-telomere field, 86
centromeres are silenced, 75
centromeres do not arise *de novo*, 76
centromere structure, 76
centrosomes, 37
Chalcolepidius, 145
change from bilateral to radial symmetry, 180
change of chromosome pattern, 155
Chargaff's rule, 20
chemical copying, 169
chemical elements, 11
chemical elements obligatory for cell construction and function, 179
chiasmata, 14, 41, 67, 75, 134
chiasmata formation, 15, 134
chicken, 137
chicken ovalbumin gene, 30
chimaeras, 32, 145
chimpanzees, 37
Chinese hamster, 81
Chlamydomonas, 30
chlorophyll, 141, 179
chloroplast DNA, 23, 117
chloroplast genome, 117
chloroplast motility, 91
chloroplast ribosomes, 117
chloroplasts, 22, 30
chloroplast structure, 117
cholera bacillus, 10
Christmas trees, 25, 85
chromatin sheets, 155
chromatin silencing, 137
chromomere, 67, 71
chromomere patterns, 15, 41, 75, 133
chromomere size gradient, 34, 75, 133, 176
chromosome as an aperiodic crystal, 22
chromosome diagnostic, 49

chromosome field, 175, 176
chromosome field concept, 34
chromosome genotype, 133
chromosome III of yeast, 36
chromosome integrity, 175
chromosome length, 183
chromosome missegregation, 68
chromosome models, 169
chromosome movements, 95
chromosome numbers, 145
chromosome of T2, 105
chromosome pairing, 68
chromosome pattern, 133
chromosome pellicle, 169
chromosome periodicity, 146
chromosome phenotype, 133
chromosome prehistory, 2
chromosome rearrangements, 18
chromosome replication or separation, 77
chromosome rings, 15, 151
chromosome scaffold, 170
chromosome theory of heredity, 14, 71
chromosomes, 11
chromosomes as hereditary determinants, 14
chromosomes move constantly in the nuclear sap, 77
chromosomes with a star configuration, 41
chromosomes with two centromeres, 77
chronic myelogenous leukemia, 151
cilia, 95
ciliates, 113
ciliated protozoa, 82
circular chromosome, 109
cisternae, 97
cloned DNAs, 28
cloning of humans, 1
cloning of the sheep Dolly, 37
clustering of genes with similar functions, 137
clustering of housekeeping genes, 137
cobalt, 179
Coccids, 75
codon recognition, 40
coenzymes, 141
cohesive ends, 28
colchicine, 47, 159
colchicine induced polyploidy, 18
cold induced regions, 134
colloidal emulsion, 99
Commelinaceae, 146
comparative anatomy, 6
composition of ancestral chromosomes, 141
compound microscope, 7
concept of the chromosome, 6
conjugation process in bacteria, 23
conservatism, 137
conserved domains, 34
constancy of distance between chromomeres, 134
constrictions, 85
control mechanism, 162
convergence, 145
core histones, 41
cows, 127
Crepis, 157
cricket *Acheta*, 155
Cricotopus silvestris, 145
crosses in animals, 14
crossing over, 18, 29, 37, 41, 67, 75, 137, 151
crown gall bacterium, 28
crystallography, 22
crystals, 179
cut-and-patch repair, 183

cyanobacteria (blue-green algae), 30
cyclic AMP, 32
cystic fibrosis, 32
cystic fibrosis genes, 37
cytochrome c, 137, 141
cytochrome c method, 127
cytoplasmic inheritance, 14
cytoplasmic movements, 91
cytoskeleton, 59, 91

Datura, 15, 159
deer species, 86, 146
defective mutations, 183
deficiency-duplication cycle, 151
delimitation of the centromere, 77
de novo telomere addition, 82
density gradient equilibrium ultracentrifugation, 23
development, 137
diamond knife, 22
diastase, 7
different amino acid sequences, 146
different parts of a flower, 138
different stages of mitosis, 65
differential formation of bands, 155
diffuse centromeres, 75
diploid number, 67
directed mutations, 43
discovery in science, 6
discovery of mutations, 11
diseases, 141, 151
disintegration at pro-metaphase, 86
dispersed TY1/copia, 170
distance telomere-centromere, 134
distorter genes, 68
distribution of the ribosomal genes, 86
division cycle, 170
division of human leukocytes, 24
DNA, 47
DNA-DNA hybridization, 146
DNA finger print, 32
DNA in cell organelles, 18
DNA in chloroplasts, 23, 63
DNA in mitochondria, 23, 63
DNA ligase, 141
DNA model, 23
DNA of the centromere, 41
DNA polymerase, 29, 141
DNA proofreading, 141
DNA recombination, 175
DNA repair, 37, 141
DNA replication, 77, 175
DNA-RNA hybridization, 25
DNA-RNA hybrids, 28
DNA satellites, 133, 161
DNA sequencing, 2
DNA synthesis, 23
DNA technology, 24
Down's syndrome, 24
Drosophila, 14, 19, 24, 25, 29, 33, 34, 37, 47, 68, 77, 81, 85, 109, 151, 159, 162, 175
Drosophila genome, 2
Drosophila melanogaster, 155, 161
drugs, 30
duration of the cell cycle, 67

E. coli, 23, 28, 30, 37, 109, 117, 137
E. coli genome sequence, 2
effect at distance, 40
electron microscope, 18, 48
electrons, 123

electrophoresis, 18, 49
electrostatic forces, 76
elementary particles, 123
embedding in paraffin, 47
embryology, 14
embryonic chimaeras, 63
embryonic development, 11
embryonic field, 34
emergence of the eukaryotic chromosome, 127
Encyclopaedists, 6
endocytosis, 59, 99
endoplasmic reticulum, 22, 97
endoreduplication, 128, 159, 169
endosomes, 59
enucleated frog's eggs, 25
enzyme triose phosphate isomerase, 137
Ephestia, 19
episomes, 109
ergastoplasm, 97
error-prone repair, 183
Euglena, 128
Euglena gracilis, 30, 63, 113, 117
Euplotes, 81
evolutionary histories, 183
evolution of the chromosome, 183
evolution of the protozoan chromosome, 128
exceptions to the code, 113
excision and elimination of DNA segments, 128
exocytosis, 99
exons, 30, 42
extrachromosomal gene inheritance, 63

fate of cancer cells, 63
ferredoxin, 137
fertilization, 11, 67
Feulgen test, 76, 117
fibroin gene, 29
field of action, 42, 175, 183
first enzyme, 7
first photographs, 7
five-fold symmetry of flowers, 179
fixatives, 47
flagella, 95
flanking regions, 162
flanking sequences, 137
flowering plants, 175
fluorescent dyes, 49, 63
formed at fixed positions, 134
fragile site of the human X chromosome, 36
fragile sites, 162
fragile X syndrome, 162
fragments, 151
Fritillaria, 134
fungi, 142
fusion of duplicated genes, 162
fusion of the embryos, 48
Fusomes, 37

galactosemia, 28
galactose operon, 28
Gaucher's disease, 32
gene, 1, 14, 15
gene activity, 25
gene amplification, 19, 24, 128, 161
gene blocks, 141
gene clusters, 33, 175
gene conservation, 137
gene dense regions, 175
gene diversification, 43
gene effect extended over 100 bands, 175

gene expression, 40, 42
gene for an alanine transfer RNA, 28
genes for 18S ribosomal RNA, 175
genes for 28S ribosomal RNA, 175
genes for 5S RNA, 85
genes for ribosomal RNA, 161
genes for ribosomal proteins, 175
gene isolation, 162
gene libraries, 29
gene light regions, 175
gene number, 183
gene permanence, 43
gene polarity, 175
gene synteny, 141
gene territory, 41, 86, 175
gene therapy, 37
genes are silenced, 159
genes at the centromere, 76
genes involved in memory, 37
genes with related functions were clustered, 137
genetic code, 23, 24, 30, 113, 128, 180
genetic linkage map of human genome, 2
genetic linkage maps, 32
genetic map of *Drosophila*, 14
genetic music, 183
genetic noise, 183
genetic recombination, 19, 109
genetic transformation, 19
genetics, 14
genetics of eye pigments, 109
genome imprinting, 163
germ cells, 67
germinal vesicle, 157
germ warfare, 19
giant chromosomes, 47
giant neurons, 32
giant suspensor chromosomes, 157
glass knives, 22, 48
globin gene families, 137
globin genes, 162
glucocerebrosidase gene, 32
glycoprotein, 184
Golgi apparatus, 22, 97
gonial metaphase, 155
gorillas, 37
gradients and fields in chromosomes, 34, 176
gradients of molecules, 34
Gramineae, 81, 146
grana, 117
grasshopper, 67
guidance of crossing over, 37

H3 uridine, 25
haemoglobin, 25, 141, 179
Haemophilus influenzae, 36, 109
hand centrifuges, 48
Haplopappus, 29, 145, 146
Haplopappus gracilis, 145
harbor seal, 141
harlequin chromosomes, 28
hemicellulose molecules, 99
herbicides, 34
heterochromatic regions, 23, 145
heterochromatic regions do not occur at random, 133
heterochromatin, 15, 23, 42, 75, 82, 133, 175
higher mammals, 86
high-mobility proteins, 133
H. influenzae first bacteria sequence, 2
histochemistry, 7
histone IV, 141

histone genes, 29, 161
histone proteins, 133
histones, 30, 40, 47, 127
Holomastigotoides, 128
Holomastigotoides tusitala, 127
Holosticha sp., 127
homeobox gene, 34
homeobox gene family, 138
homeobox protein, 36
homeotic genes, 33
Homo sapiens, 81
honey bees, 145
horse, 141
horse-donkey hybrids, 30
house-keeping functions, 42
human, 137, 141, 146, 175
human anti-oncogenes, 36
human cells fused with carrot protoplasts, 29
human genes, 24
human gene therapy, 32
human genome, 184
human hybridomas, 28
human mitochondria, 30
human oncogenes, 1
human triploid individuals, 159
human vertebral column, 138
human X chromosome, 137
humans, 37, 75, 81, 86, 146, 175
Huntington's disease gene, 36
hybridization *in situ*, 24, 49
hyperacetylation of histones, 40
hypoacetylation, 40
hypotrichous ciliates, 128

Ichthyomys pitteri water rat, 146
immune system, 184
immunoglobulin chain, 30
imprinting an address, 97
increase of DNA content, 25
independent assortment of characters, 14
Indian and Chinese muntjacs, 81
Indian muntjac, 141
inducer, 24
informational families, 161
informational RNAs and proteins, 67
inherent mechanisms of surveillance, 183
inheritance in plants, 11
inheritance of acquired characters, 11
inhibitory effects, 75
inosine, 40
insect, 137
in situ analysis of DNA, 25
in situ hybridization, 33, 162
insulin, 28
interaction between centromeres, 77
interaction between genes, 175
interactions affecting gene expression, 175
interactions between specialized chromosome regions, 86
interferon, 28
interphase, 23, 71
interphase nucleus, 65
interstitial telomeres, 75, 81
introns, 30, 42
inversions, 151
in vitro enzymatic synthesis, 23
iodine, 7
iron, 179
iso-chromosomes, 20, 76

Kaiser-Wilhelm Institutes, 22

kilogram, 6
kinetochore, 76
kinetoplast, 113
Klinefelter's syndrome, 24
knobs, 71, 85, 134
knob sequences, 183
KU, 81

Labiatae, 81
lac operon, 24, 109
lambda, 137
lambda viruses, 24, 28
lampbrush bodies in oocytes, 11
lampbrush stage, 71
lampbrush structure, 67
large blocks of heterochromatin, 146
large field binocular microscope, 47
later DNA synthesis at meiosis, 29
late replicating heterochromatic regions, 23
law of octaves, 11
Leber's disease, 32
left-handed and right-handed, 123
left-handed and right-handed forms, 180
left-handed and right-handed symmetry, 179
Leguminosae, 157
length of the centromere-telomere region, 184
length-sensing mechanism, 183
leptin receptor, 37
Liliaceae, 81, 183
linkage, 14
linker histones, 41
liquid emulsion, 48
liverwort, 63
living cells, 47
lizard, 75
localize DNA, 15
location of centromeres, 76
location of genes, 18
location of genes in human chromosomes, 28
long DNA synthesis, 67
long interspersed elements (LINEs), 170
long-range effect, 42
long-range mechanisms, 163
Luzula, 75
lysosome, 22
lysozyme, 25

magnesium, 179
magnification, 161
maintenance of gene order, 138
maintenance of harmony, 183
maintenance of the same chromosome number, 65
maize, 18, 75, 81, 137, 151, 183
mammalian viruses, 105
mammals, 33
Marchantia polymorpha, 32, 117
Marfan's syndrome, 36
Marshall plan, 22
mass isolation of chromosomes, 47
maternally derived genes, 163
mathematical analysis of population genetics, 15
matrix, 169
Max-Planck *Abteilung*, 22
meiosis, 67
mei-S332 gene of *Drosophila*, 37
Melanoplus, 23
memory at the molecular level, 32
memory genes, 32
Mendel's laws, 11, 14
mental retardation, 36, 162

Mesocyclops edax crustacean, 146
messenger RNA, 23
metabolic DNA, 19
metacentrics, 76
metazoan evolution, 128
meter, 6
micellae, 99
microbial genetics, 19, 109
microbiology, 22
microfilaments, 59, 91
microsatellites, 42, 161
microscope, 6
microtubule proteins, 95
microtubules, 37, 59, 95, 128
mimetic mutants of tomato, 137
minerals, 179
mini-chromosomes, 42
minisatellites, 42, 161
minor spliceosome, 42
mismatch repair, 183
mitochondria, 22, 113
mitochondrial genes, 113
mitochondrial genome, 32
mitosis, 67
mitotic-like machinery, 109
model of DNA, 22, 169
modification of transfer, ribosomal and messenger RNA, 42
molecular mid-wives, 33
molecular pathology, 113
molecular transit, 93
monitoring of messenger RNAs, 141
moulds, 19
mouse, 33, 34, 63, 137, 141, 146, 161, 175
mouse cells, 30
mouse embryonic germ cells, 32
movement of chromosome ends, 81
multiple nature of the telomere, 81
multiple translocations, 145
multiplicational families, 161
Muntiacus muntjak deer, 146, 184
Muntiacus reevesi, 146
murine leukemia virus, 33
mutation, 137
Mycoplasma genitalium, 36
myoglobin, 25
Myrmecia ant, 146, 184
Myrmecia pilosula, 145

nascent RNA molecules, 25
nematodes, 142
neo-Darwinism, 11
nerve cells, 12
nerve ganglia cells, 155
neural processes, 32
Neurospora, 19, 109, 175
neutrino, 123, 180
new RNAs U4 and U6 atac, 42
nitrocellulose filters, 28
no DNA synthesis at interphase II, 67
non-disjunction, 15
nonhistone proteins, 170
nonrandomly distributed transposable elements, 151
northern blotting, 29
novel chromosome combinations, 67
nuclear envelope, 59, 93, 128
nuclear membrane, 65
nuclear RNAs, 42
nuclear sap, 41, 93
nuclear transplantations, 25, 63
nucleic acid polynucleotide phosphorylase, 23

nucleolus, 18, 59
nucleolus-less mutant, 162
nucleolus organizer, 18, 24
nucleosome core, 77
nucleosomes, 30, 40, 170
number of chromatids, 169
number of human primary genes, 183
number of introns, 43
number of proteins, 183

obesity in the mouse, 36
octameric histone core, 133
Oenothera, 15, 146, 151
oil immersion objectives, 47
Onagraceae, 146
oncogenes, 123, 137
one gene one enzyme hypothesis, 19, 109
onion, 65
onset of cancer, 162
operator genes, 24
operator sequence, 141
operons, 24, 105, 137
Ophioglossum fern, 159
optimal centromere-telomere length, 183, 184
optimal territory, 184
orangutans, 37
order of genes in bacteria and viruses, 137
ordered mutation, 34
ordered process in the production of mutations, 30
origin and evolution of the genetic code, 138
origin of eukaryotic organisms, 127
origin of life, 10, 179
origin of the cell, 179
origin of the chromosome, 179
origin of the nucleolus, 85
origin of split genes, 137
origin of viruses, 105
Ornithogalum virens, 71, 134

P32, 48
pachytene, 42
pachytene checkpoint, 68, 86
pachytene stage, 71
packaging of chromosomes, 67
packing of the DNA, 170
pairing, 67
palindromes, 48
Papaver rhoeas, 157
Paramecium, 81
Paris, 134
paternally derived genes, 163
pathologic conditions, 159
peas, 127
pectin, 99
Periodic Table, 179
permanent cell organelles, 11
peroxisomes, 59
phage genetics, 19
phage MS2, 28
phagocytes, 10
phase contrast microscope, 18, 47
phaseolin gene, 33
Phaseolus vulgaris, 157
phlogiston, 7
phosphate groups of the DNA, 40
phosphorus 32, 22
photo-reactivating enzyme system, 183
photosynthesis, 63, 117
phytohemagglutinin, 24
picro-carmin technique, 11

pig, 141
pigment synthesis, 19
plant-human cytoplasm, 29
plants, 142, 146
plasmacytomas, 30
plasma membrane, 59
plasmid-linked genes, 33
plasmids, 28, 109
ploidy level, 117
point mutations, 162
polarity in the DNA molecule, 175
polarity of gene transcription, 137
polar mutations, 175
polonium, 12
polycentric chromosomes, 76
polyethylene glycol, 29
polymerase chain reaction (PCR), 33, 49
polyphenylalanine, 23
polyploids, 47, 128, 159
polyteny, 128
polytene structures, 18
position effects, 161, 175
positron, 123
pregnancy, 49
premature condensation of the chromosomes, 29
prenatal test, 28
pre-ribosomal RNA, 85
pretreatments of cells, 24
primary function of mitochondria, 113
primary genes, 183
principle of the correlation of parts, 7
Pristiurus shark, 157
processed pseudogenes, 162
processing and modification of transfer, ribosomal and messenger RNA, 42
programmed death, 32
promoter, 24, 86
promoter sequence, 141
proofreading, 183
proofreading mechanism, 141
properties of the elements, 11
proteasomes, 59
proteinaceous cap, 42
protein-DNA recognition, 40
protein gradients, 34
protein in cell organelles, 18
protein kinases, 28, 37
protein layers, 99
protein linkers, 170
protein p53, 141
protein periodicity, 146
protein profiles, 43
protein synthesis, 23, 85
proteins, 141
proteins essential for photosynthesis, 117
proton, 123
protoplasm, 7
protozoans, 33, 42
pseudogenes, 42, 137, 162
puffs, 71
puffs of polytene chromosomes, 25
pulsed field gels, 2
pulse field gradient electrophoresis, 33
pure diploids, 159
purine bases, 141
pyruvate kinase M gene, 162

quantity of DNA in different tissues, 19
quinacrine dyes, 28

rabbit beta-globin genes, 30
rabies, 10
radial symmetry of starfishes, 179
radioactivity, 12
radio-isotopes, 22
radish (*Raphanus sativa*), 15
radium, 12
random events, 151
Raphanobrassica, 15
RAS proteins, 59
rat cells, 32
reassembly at telophase, 86
reassembly of the nuclear envelope, 93
RecA, 37
reciprocal translocations, 146, 151
recombinant DNA technology, 28
recombination of exons, 162
recombination repair, 183
reduction of the chromosome number, 67
redundancy of DNA, 161
regulatory circuits, 43
regulatory genes, 34
reiterated (or redundant) DNA, 162
repair mechanisms, 183
repair of double strand breaks (DSB), 67
repeated DNA, 161
repetitive DNA, 24
replication units, 170
repressor, 24
Research Councils, 22
resolvases, 183
resting stage, 71
restriction endonuclease *AluI*, 161
restriction enzymes, 28, 48
restriction map of SV40, 2
reticular formation of the cytoplasm, 91
retinoblastoma, 32
retrotransposition, 162
retrotransposons, 151
Rhizobium, 33
Rhoeo, 146, 151
Rhynchosciara, 25
Rhynchosciara angelae, 157
ribosomal DNA amplification, 67
ribosomal RNA, 24
ribosomal RNA genes, 175, 184
ribosomes, 23
ribozyme, 41
RNA, 47
RNA-dependent DNA polymerase, 28
RNA editing, 33
RNA in cell organelles, 18
RNA as a jack of all trades, 41
RNA primer, 41
RNA splicing, 128
RNA surveillance, 141
RNA transcription, 175
RNAs with no established role, 42
Rockefeller Foundation, 22
Rosa canina, 68
R. rubrum, 137
rotation of the Earth, 7
Rous sarcoma virus, 28
rye, 42, 75, 76, 81

salamander, 65
salivary glands, 71
same functional solution, 146
satellite DNAs, 42
scaffolding of the chromosome body, 41

scaffolding proteins, 170
S. cerevisiae sequence complete, 2
Sciara, 68
science is a fragile process, 6
search for the microbes, 10
sea urchin, 65
secondary constrictions, 85
selection, 11
self-assembly, 25, 138, 179
self-assembly mechanisms, 95
self-assembly of viruses, 25
self-catalytic process, 163
self-propagating infectious RNA, 24
self-regulation systems, 151
self-splicing, 162
semi-conservative distribution of labeled DNA, 23
sendai virus, 29
sensing mechanisms, 161, 183, 184
sensory neurons, 32
sequence of amino acids, 22, 48
sequence of human chromosome 22, 2
sequence of spiralization and despiralization, 133
sequence tag sites (STS), 2
sequencing of DNA, 29, 36, 49
sequencing of mitochondrial DNA, 37
sequencing of the bases, 48
sequencing of the genome, 2
sequencing of the human genome, 37
sequencing of the natural ends, 81
sex chromosomes, 157, 159
sex determinants, 14
short interspersed elements (SINEs), 170
silk worm, 29
silk-worm disease, 10
silver nitrate, 12
simian virus 40 (SV40), 28
simple sequence families, 161
single-copy genes, 33
single cell layer, 47
size of the chromosomes, 146
size of the gene, 19
size of the structural gene, 37
Solanaceae, 81
somatic cell genetics, 24
somatostatin, 28
SOS response, 183
Southern blots, 2
Southern method, 29
spacer regions, 86
special GG base pairs, 82
special telomeric proteins, 41
Spectrosomes, 37
speed of light, 7
spermatids, 155
spermatocytes, 71
spermatogonia, 71
S phase, 95
spindle, 37, 65
spindle pole body, 77
spindle poles, 95
spiral structure, 48, 128, 169
Spirotrichonympha polygyra, 145
splicing, 162
splicing process, 33
split genes, 30, 42, 183
spontaneous generation, 10
squash technique, 15
stacking forces, 40
staphylococcal nuclease, 146
starch, 7

Stentor, 63
stratification of cell organelles, 48
streptavidin, 33
stripping film, 48
stroma, 117
structural basic plan, 7
structural changes, 151
structural genes, 24, 43
structural proteins, 41, 133
structure of DNA, 2
structure of telomeres, 81
Stylonychia, 63
Stylonychia lemnae, 128
sub-metacentrics, 76
sunflower, 33
superfamilies, 141, 183
super-oxide dismutase, 36, 37
superwobble, 40
suppressor genes, 32
suppressors, 123
surface antigens, 184
symmetries, 179, 180
synaptonemal complexes, 67, 127, 128
syntenic genes, 141
synthesis between genetics and cytology, 12
synthesis of urea, 7

T4 phage, 105
tandem repeats, 170
Taraxacum, 159
target theory, 19
telescope, 6
telocentrics, 76
telomerase, 81, 184
telomerase activity, 36
telomerase enzyme, 41
telomere binding proteins, 81
telomere concept, 81
telomere (end region), 42
telomere length and cell aging, 82
telomere regions, 30
telomere replication, 41
telomere-telomere distance, 183
telomeres can also arise anew, 82
telomeres can function as centromeres, 82
telomeric DNA, 42
term biology, 7
term mitosis, 11
term mutation, 14
term protein, 7
terminal organizers, 183
terminal repeat sequence, 128
tetracarcinoma cells, 28
Tetrahymena, 30, 127
tetraploidy, 159
T genes of the mouse, 36
theory of biological evolution, 7
theory of the cell, 7
thermal disruption, 141
thylakoid membranes, 117
thymidine kinase, 24
tissue smear, 47
tobacco mosaic virus, 18, 25, 105
tobacco plants, 32
tomato, 81, 155
total RNA mass of a cell, 41
totipotency, 29
Tradescantia, 42
Tradescantia paludosa, 67
transcriptional silencing, 40

transcription factors, 133
transcription of the genes was polarized, 137
transcription process, 49
transfer experiments, 33
transfer of a gene, 1
transferred human genes, 37
transfer RNA, 23, 137
transformed cells, 49
transgenic lines, 36
transgenic mice, 33, 37
transit of molecules, 97
translocation between chromosomes 6 and 9 of maize, 85
transport and endocytic vesicles, 59
transposable elements, 20, 151
transposases, 183
transposons, 30, 34, 151, 183
trans-splicing of introns, 163
TRF1, 81
TRF2, 81
Trillium, 134
triploidy, 159
triploidy in humans, 128
trisomic plants, 15
trisomy, 24, 159
tritiated thymidine, 23, 170
Triticum aestivum, 67
tritium (H3), 22, 48, 97
Trypanosoma, 184
Trypanosoma brucei, 81
trypanosomes, 33, 113
trypsin, 137
tubercle bacillus, 10
tumor-inducing plasmid, 28
Turner's syndrome, 24

U11 RNA, 42
U12 RNA, 42
ultracentrifugation, 18, 23
ultracentrifuge, 15, 48
ultramicrotome, 22
ultra-violet microscope, 47
union of sister chromatids, 37

vaccination against smallpox, 10
van Beneden's laws, 11
variable genes, 184
variant of histone H2B, 81
variation in chromosome number, 145
variation in size, 157
vertebrates, 142
very long prophase, 67
very low chromosome numbers, 145
Vicia faba, 23
viral chromosomes, 105
viruses, 105
virus T4, 23
vitamin B12, 179

wall of plant cells, 99
Walterianella, 157
water content, 40
water in cells, 40
wheat, 75, 137
whole arm transfer, 141
wings of insects, 138

X chromosome, 71
X chromosome of *Drosophila melanogaster*, 155
X chromosomes of the mammalian female, 24
Xenopus, 24, 63, 67

Xenopus laevis, 162
X-ray analysis, 99
X-ray analysis of DNA, 22
X-ray diffraction, 48
X-ray photographs of proteins, 22
X-rays, 12, 15, 30, 81

Y chromosome, 151
Y chromosome of D. hydei, 67

yeast, 19, 24, 28, 77, 82, 86, 127
yeast artificial chromosome (YAC), 2
yeast chromosome XI, 36
yeast mitochondria, 30
yeast protein RAP1, 123

zebra fish, 36
zinc, 179
zinc fingers, 179